Value Analysis and Engineering Reengineered

Value Analysis
and Engineering
Reengineered

The Blueprint for Achieving Operational Excellence and Developing Problem Solvers and Innovators

Abate O. Kassa

CRC Press
Taylor & Francis Group
Boca Raton London NewYork

CRC Press is an imprint of the
Taylor & Francis Group, an **informa** business

A PRODUCTIVITY PRESS BOOK

CRC Press
Taylor & Francis Group
6000 Broken Sound Parkway NW, Suite 300
Boca Raton, FL 33487-2742

© 2016 by Abate O. Kassa
CRC Press is an imprint of Taylor & Francis Group, an Informa business

No claim to original U.S. Government works

Printed on acid-free paper
Version Date: 20151012

International Standard Book Number-13: 978-1-4987-3725-8 (Hardback)

Visit the Taylor & Francis Web site at
http://www.taylorandfrancis.com

and the CRC Press Web site at
http://www.crcpress.com

To my beloved wife, Waka, and the amazing

long journey we traveled together,

my remarkable and loving sons, Mengistu and Alem,

my precious granddaughter, Sofia, and her

wonderful and beautiful mother, Paola.

Contents

SECTION II   The Value Methodology

SECTION III   The Value Organization

Foreword

I have had the privilege of knowing Abate Kassa for over 30 years. We first met while he was executive director of ISM-New York, local affiliate of the Institute for Supply Management, where he served with distinction for 19 years. I had become an active volunteer with ISM-New York as a committee member, then a member of the board of directors and, later, president of the affiliate. Working closely with Abate Kassa through all of those years, he quickly became my mentor, my teacher, and a trusted advisor.

As executive director of ISM-New York, Abate Kassa was our resident expert on Value Analysis/Value Engineering (VA/VE). In this role, he was focused on customer needs and cost discipline. Driven by the *better for less* ethos of VA/VE, he transformed the affiliate into a center of excellence in professional development for procurement and supply professionals, thereby earning the prestigious Leonard Award, ISM-New York's highest honor recognizing outstanding contributions to the affiliate and to the procurement and supply management profession. Furthermore, Abate Kassa has been conducting seminars on VA/VE through all of these years, with his presentations routinely receiving "Excellent" ratings from participants, setting a high standard for other instructors. In addition to presenting for ISM-New York, Abate Kassa has literally travelled across the globe presenting for organizations of every variety, from professional associations to corporations, government entities, and nonprofit organizations, helping them all learn to achieve operational excellence by adopting VA/VE as their corporate DNA. In many of his endeavors, he oversaw real-life initiatives to help his students actually implement what he taught them. Abate Kassa does not just talk about VA/VE, but he lives it!

You may be wondering, is this just another book on VA/VE? After all, VA/VE has been around for over 50 years and there have been other books on the subject. Abate Kassa has updated and upgraded VA/VE, thereby enhancing its value; and it is my pleasure to introduce you to this important work. What distinguishes Abate Kassa's *Value Analysis & Value Engineering Reengineered* is that he connects the dots of a variety of management tools such as Lean Production, Six Sigma, Total Quality Management, Kaizen, Business Process Reengineering, and Project

Management by integrating them into the *value methodology* he has dubbed "PISERIA."

Over the years, I have seen many organizations grab on to the "program of the month." For example, they may see that successful companies are "doing Six Sigma," so they decide to give it a try. They hire a consultant, form a team, and pick a project, but they don't really get behind it and it doesn't become ingrained as part of the culture. When results fall short of expectations, they drop it and move on to the next "program." In addition, they may not have chosen the proper tool for the task at hand. Paul Hersey and Ken Blanchard describe how leaders need to apply "situational leadership" where, rather than having one leadership style, leaders need to adapt their leadership style to the goals of the project and capabilities of the individual team member. In a similar way, Abate Kassa provides a tool kit and shows us how to apply the proper management tool to the task at hand. With Abate Kassa's reengineered VA/VE set forth in this book, you will learn how to take the value methodology from a project or a program, to a way of thinking, a way of doing business.

Martin J. Carrara
Senior Corporate Counsel
Pfizer, Inc.
New York City, New York

Acknowledgments

This book is a product of my dual lenses of experience and research over many years, but it would not have been published without the collaboration of my good friend James Martin of ISM-New York, who introduced me to Tom Cook of Blue Tiger International, who in turn recommended me to the right publisher. I thank them most sincerely for their undivided support.

Introduction: Making the Case for Value Analysis and Value Engineering

Successful companies start with a deep understanding of customer needs. They listen to the voice of the customer and determine the customer's critical-to-quality (CTQ) requirements and convert those into specific product or service design features that can meet or exceed the customer's needs. CTQ elements are the attributes and expectations most important to the customer. As the great thought leader Peter Drucker taught us, the right questions to ask are, "Who are our customers?" and "What does the customer value?"* Without a customer, we have no business.

Yes, everything begins and ends with the customer because the purpose of business is to create and keep a customer. Customers rightfully expect the best quality at the lowest possible cost (value entitlement). Organizations are expected to be agile and produce or provide customer-perceived value for products and services. Customers control the votes on satisfaction and value. Focusing on customer-perceived value is the essence of *value analysis (VA), value engineering (VE), value management (VM),* and *value thinking.* Since their objectives and methodology are the same, these terms can be used interchangeably. However, when we make distinctions between them, we say that value analysis is to existing products as value engineering is to new products. Since Section II of this book is entirely devoted to the value methodology and it is briefly discussed in Chapter 1, it will suffice to mention here that value methodology is the "operating system" of VA/VE, to use a computer metaphor.

Furthermore, as illustrated in Figure I.1, VE is applied at the design stage while VA is applied after commercialization and customer feedback. As value management evolved and was applied to larger and more complex products and systems, emphasis shifted to "upstream" product development activities, where the value methodology can be more

---

* *The Five Most Important Questions You Will Ever Ask About Your Organization* (San Francisco, CA: Jossey-Bass, 2008).

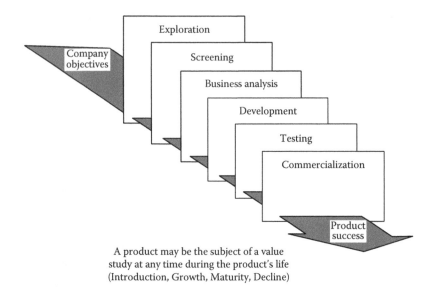

A product may be the subject of a value
study at any time during the product's life
(Introduction, Growth, Maturity, Decline)

FIGURE I.1
Stage of product evolution.

effectively applied to a product before it reaches the production (manufacturing) phase.

As reported by Mikel Harry and Richard Schroeder in their book *Six
Sigma** and diagrammed in Figure I.2, while design typically represents
the smallest actual cost element in products, it leverages the largest cost
influence. Design determines 70% of a product's total cost. An incremental improvement in the design has a huge direct impact on cost. For example, a 30% savings through design simplification would translate into over

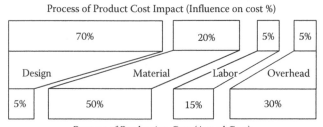

High Leverage of Process and Product Design on Cost

FIGURE I.2
Six Sigma leverage in product design.

---

* *Six Sigma: The Breakthrough Management Strategy Revolutionizing the World's Top Corporations*
  (New York: Currency, 2005).

21% overall cost savings. An overwhelming majority of defects are created during the design process. There is a tremendous premium on getting it right and a tremendous penalty for getting it wrong. Therefore, try to do it right the first time, all the time, with no excuses.

Value analysis is the parent discipline while value engineering is value analysis applied at the design stage. In the vernacular of management, value analysis is management by correction (a posteriori) and value engineering is management by prevention (a priori).

My favorite example to illustrate the difference between VA and VE comes from the commercial aviation industry, where I spent nearly two decades of my working life. The case is that of American Airlines and Southwest Airlines. American Airlines had fourteen different types of airplanes and consequently had difficulty meeting CTQ needs for on-time departures. Furthermore, problems with training mechanics to maintain fourteen different airplanes and an inventory of noninterchangeable parts were burdensome. American Airlines correctly determined that the way to build resilience was to reduce complexity. Accordingly, the airline reduced its fourteen types of airplanes to seven (a case of VA). Unfortunately that decision came too late as AMR Corporation, the parent company of American Airlines, filed for Chapter 11 bankruptcy protection in 2011. (Two years later, American Airlines and U.S. Airways merged on December 9, 2013.) On the other hand, Southwest Airlines made the right strategic decision from the beginning by killing complexity and standardizing on only one type of aircraft (a case of VE), the Boeing 737, which happens to be the most popular commercial airplane in the world. Southwest takes pride in the fact that it marked 2012 as its fortieth consecutive year of profitability—forty years in the black!* That is a record unmatched in the airline industry, and the trend continues.

Figure I.3 represents the "on the one hand and on the other hand" formulation. With VA/VE, we are expected to perform a unique balancing act by mastering the delicate balance between achieving the best quality of product or service at the least cost in resources that leads to an optimization nirvana. As it is commonly understood, trying to do the best at the least cost in resources is a search for excellence—hence the theme of this book, VA/VE as the blueprint for achieving operational excellence. In other words, what we are trying to achieve is the *best value for the money.*

---

* Aaron Karp, Southwest earns 40th straight annual profit in 2012, *Air Transport World* (January 24, 2013), accessed August 31, 2013, http://atwonline.com/airline-finance-data/news/southwest-earns-40th-straight-annual-profit-2012-0124.

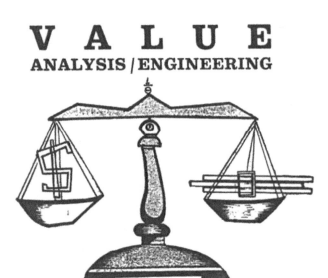

FIGURE 1.3
The balancing act.

In the face of the high cost of doing business, over the years organizations have taken various cost-cutting initiatives such as implementing cost-reduction programs and austerity programs, with each approach yielding results that did not meet the desired objectives. "Cost reduction" is not an attractive term—nobody wants to be reduced. In view of people's need for security, a cost-reduction program is inevitably frustrating, causing fears about loss of jobs, earnings, status, and so forth. This may then result in complete blocking of the cost-reduction program. Moreover, unbalanced focus on cost reduction kills innovation and quality suffers, leading to a mediocrity trap. Cost is a design parameter because it controls resource expenditure. Spending is what is done to create a desired outcome. Wherever cost is incurred, there is an opportunity for VA/VE. Therefore, costs are obviously important. But we should be value-thinking about "cost management" instead of "cost cutting." Cost cutting is a short-term solution. Cost management focuses on improving overall management practices by such means as embracing lean thinking, shortening the total lead time, employing minimum resources, improving quality, and reducing overall cost of operations. Merely proposing that a cost-reduction program is needed implies that the organization hasn't been doing very well. It portends an era of austerity.

Austerity programs, also known as crash programs or belt-tightening programs (or sequestration in the case of the U.S. government), are signs of mismanagement. Organizations that follow this path have failed to appreciate the simple truth that the best producer is also the lowest cost provider. Executives who issue arbitrary cost-cutting directives deceive themselves if they think the savings will last any length of time. Typically, the result is loss of employee confidence in management, poor employee morale, poor-quality work, and lower productivity followed by the development of a protective layer of fat in the form of excess or redundant manpower as an insurance against the next round of mandated reductions. This karate chop of austerity programs is an expensive means of cost reduction as it could lead to a loss of customers and market share and experienced staff. Companies that have resorted to such economy drives have learned that this sort of unskilled surgery all too often cuts away two pounds of organizational muscle with every pound of corporate fat. It would be helpful to remember the maxim, "If all of us could learn to live without excesses today, we might not have to live without essentials tomorrow."

By contrast, and to stave off the need for either of these traditional and reactive approaches, I offer VA/VE/VM as a disciplined and effective methodology that can help businesses develop a workforce of problem solvers and innovators that will enable them to achieve operational excellence and a sustainable competitive advantage. Here are some more arguments for the value of VA/VE:

- VA/VE is a parallel-thinking methodology that results in concurrent improvements in performance and reliability and the elimination of unnecessary costs for products and systems of operations.
- VA/VE provides a roadmap for achieving operational excellence and a toolkit that can help foster a *value-for-money* (VfM) culture.
- The practice of VA/VE unleashes the potential of employee creativity and innovation through helping them find meaning in work by providing them with the opportunity for self-actualization and achievement, thereby helping them make a difference. With the value improvement process, everyone benefits.
- In the world of VA/VE, we identify and eliminate unnecessary work, not people. People are not costs to be cut, but assets to be developed or redeployed. Besides, you can't shrink to greatness nor cost-save to prosperity; you achieve greatness by improving customer satisfaction and generating more profits. People are the source of the

value-creating capacity—creating value for investors, customers, suppliers, and employees. We need to heed W. Edwards Deming's Point No. 8 of his Fourteen Points of Quality: "Drive out fear, so that everyone may work effectively for the company. It is necessary for better quality and productivity that people feel secure."* Without human intervention, there can be no process improvement, hence the great value of the employee-suggestion system, in which employees are value-trained and empowered to apply VA/VE.

- Value sensitivity is the integral element of every successful operation and perpetuates itself as a way of life. It is the necessary ingredient to make organizations productive of innovations rather than resistant to them.

- **Value Analysis Lashes Unnecessary Expenses**, but a value-improvement process that attacks unnecessary costs is only doing half the job. Sure, cost effectiveness is important—in fact, it is one of the best ways to increase an organization's profitability. Actually, cost-cutting efforts (in both cost reduction and cost avoidance) often lead to increases in profit that are far greater than the increases that could be reasonably expected from sales growth. However, a truly effective value-improvement program will not only reduce costs, but also improve operations and product performance. Hence, VA/VE is a system designed to improve the bottom line (desired results) of the organization either through savings resulting from boosting sales by increasing customer satisfaction or from reduced costs.

- VA/VE is a powerful result-oriented tool for productivity improvement and innovation. It is about building capacity in critical thinking and problem solving. It is the path to entrepreneurial excellence.

- Many years of experience by business organizations applying VA/VE testify to the significant financial impact of the successful application of value methodology. In government, for instance, successful VE practices are reported by agencies such as the Federal Highway Administration (FHWA) of the U.S. Department of Transportation (DOT), for which VE studies show an average return on investment of 192:1 over the period of FY2004–2009. Table I.1 shows a summary of the FHWA's VE savings over the period 2007–2011. Therefore, value-improvement efforts (including value training) are investments rather than a cost. Accordingly, Public Law 104–106, Section 4306—Value Engineering for Executive Agencies, requires

---

* *Out of the Crisis* (Cambridge, MA: MIT Press, 2000), p. 23.

TABLE I.1

Federal Highway Administration's VE Savings, 2007–2011

Summary Of Past VE Savings for Federal-Aid and Federal Lands Highway Programs					
	FY 2011	**FY 2010**	**FY 2009**	**FY 2008**	**FY 2007**
Number of VE studies	378	402	427	388	316
Cost of VE studies plus administrative costs	$12.5 M	$13.6 M	$17.08 M	$12.47 M	$12.54 M
Estimated construction cost of projects studied	$32.3 B	$34.2 B	$29.16 B	$29.93 B	$24.81 B
Total no. of proposed recommendations	2950	3049	3297	3022	2861
Total value of proposed recommendations	$2.94 B	$4.35 B	$4.16 B	$6.58 B	$4.60 B
No. of approved recommendations	1224	1315	1460	1323	1233
Value of approved recommendations	$1.01 B	$1.98 B	$1.70 B	$2.53 B	$1.97 B
Return on investment	80:1	146:1	99:1	203:1	157:1

Source: Value Engineering, U.S. Department of Transportation, Federal Highway Administration, Washington, DC, http://www.fhwa.dot.gov/ve, accessed August 31, 2013. Updated October 18, 2012.

each federal agency to establish and maintain cost-effective value-management procedures and processes. OMB Circular A-131, Value Engineering, requires federal agencies to use value management for acquisitions and programs worth more than $1 million. Lower thresholds may be established at agency discretion for projects having a major impact on agency operations.

- The distinct characteristics of VA/VE are that it follows the scientific method adopted by value methodology; it improves value by studying the function rather than the structure of a product; and, it deliberately stimulates creativity.
- VA/VE is a proven approach by which organizations can move from the current state through a transition state to the desired future state to achieve operational excellence. The change could be either incremental or transformational.
- VA/VE is about doing work *better, faster,* and *leaner* than the competition because customer satisfaction and business success depend on the quality, speed, and cost of the goods produced and the services provided.

Yes, in times of uncertainty and harsh economic conditions, it is instinctive practice for organizations to go for a cut-and-slash approach to cost reduction, but panic-induced austerity measures can be counterproductive. For a viable future, organizations must think about their mission and strategy, not just the bottom line. Therefore, if you are passionate about excellence and wish to create, add, or deliver value, the VA/VE strategy will enable you to achieve operational excellence by increasing competitiveness through innovation and productivity, improving product performance and customer service, and eliminating unnecessary cost, waste, and inefficiency. Then the financial success of your organization could correlate directly with its operational excellence.

What is unique about the reengineered VA/VE that I present in this book is that it integrates the principles of value management with Six Sigma, Lean Thinking, Project Management, Business Process Reengineering (BPR), Kaizen, Total Quality Management (TQM), and Quality Function Deployment (QFD) by linking them together through the value methodology dubbed PISERIA based on whole-system thinking, leading to optimization. By blending these processes, VA/VE takes their power to the next level by reducing complexity and enabling you to meet or exceed customer requirements more efficiently and effectively. That is the power of simplicity of leveraging all the change-management tools into the enlightened common process of the scientific method of the value methodology. As Leonardo da Vinci said, "Simplicity is the ultimate sophistication." To sum up, the reengineered VA/VE is metaphorically like a Swiss army knife that combines several individual functions in a single unit.

Section I

The Value Perspectives

Perspective is the way you look at the future and the problems that you are trying to solve. Your perspective determines your destiny.

—Jeremy Gutsche, CEO of TrendHunter.com

THE BLIND MEN AND THE ELEPHANT: THE THEORY OF DIFFERENT PERSPECTIVES

A version of the story says that six blind men were asked to determine what an elephant looked like by feeling different parts of the elephant's body. The blind man who feels a leg says the elephant is like a pillar; the one who feels the tail says the elephant is like a rope; the one who feels the trunk says the elephant is like a tree branch; the one who feels the ear says the elephant is like a hand fan; the one who feels the belly says the elephant is like a wall; and the one who feels the tusk says the elephant is like a solid pipe.

Likewise, Section I of this book deals with an array of diverse perspectives about VA/VE (Chapters 1 through 5). Section II will be devoted to the value methodology (Chapters 6 through 13) and Section III will cover how to go about organizing a value program (Chapters 14 through 20).

1

Genesis, Goals, and Definitions

Value Analysis/Value Engineering (VA/VE) was born in 1947 at the General Electric Company in Schenectady, New York. **Harry Erlicher**, vice president of purchasing for GE, noticed at one point during World War II that when GE faced a scarcity of strategic materials, (a) a great many substitutions of materials had to be made due to the shortages and (b) many of the substitute materials resulted in not only lower cost, but also in an improved product.

As Mr. Erlicher himself put it, "This happened so often by accident, we decided to try to make it happen on purpose." To this task he assigned Larry D. Miles, who through his dedication and contributions to the science of value became known as "the father of value analysis." He also served as the first president of the Society of American Value Engineers, established in 1959 in Washington, DC, to advance the new management tool.

Larry Miles's original system was a seven-step procedure that he called the Value Job Plan. I have updated and upgraded VA's "operating system" by expanding and integrating the principles of VA/VE with Six Sigma, Lean Thinking, BPR, Kaizen, TQM, QFD, SCM, and Project Management to apply a whole-system thinking approach and created the acronym **PISERIA** (Table 1.1), which is deciphered in Figure 1.1, for the value methodology.

VA/VE is an old idea whose time has come to help us better manage today's dysfunctional institutions in business and government. The fact that VA/VE was born in 1947 does not make it out of date. VA/VE is not a fad or a flavor-of-the-month initiative. On the contrary, VA/VE is timeless and still relevant as it is based on a sound foundation of universal scientific principles. It becomes even stronger by amalgamating the new improvement methods like Six Sigma, Lean Thinking, BPR, Kaizen, and TQM with the scientific method of PISERIA.

Figure 1.1 shows the steps of the PISERIA value methodology. The details of the methodology are covered in Section II of the book.

TABLE 1.1

Traditional VA/VE vs. 21st Century VA/VE

Larry Miles's Value Job Plan*	AOK's PISERIA
1. Orientation	1. **P**reparation
2. Information	2. **I**nformation
3. Speculation	3. **S**peculation
4. Analysis	4. **E**valuation
5. Program Planning	5. **R**ecommendation
6. Program Execution	6. **I**mplementation
7. Summary & Conclusion	7. **A**udit Results

*Lawrence D. Miles, *Techniques of Value Analysis and Engineering* (McGraw-Hill, 1961), 25–28.

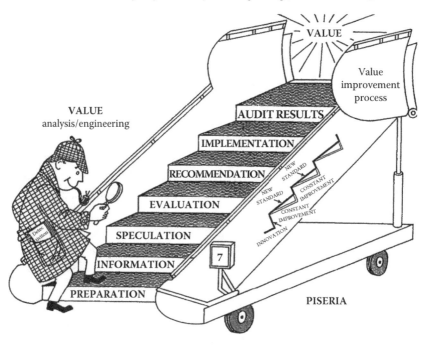

FIGURE 1.1
PISERIA.

The goals of VA/VE are performance improvement and cost-effectiveness, done in relation to each other. It is a parallel-thinking methodology, a double-edged sword. This is how you are measured in the market place: you have either performance equality or leadership, or you have cost equality or leadership. Cost leadership means that you are a lower-cost producer.

This is a "customer value proposition" that gives you a sustainable competitive advantage. A customer value proposition describes customer expectations of benefits offered and prices charged (or other value transfer). In simple words, value proposition is what the customer gets for what the customer pays. The product or service you provide is not the solution for the customer. The solution is the *function* that the product or service performs and by doing so meets or exceeds the customer's needs; it is the end result, the value the customer derives from the product or service. Hence, people do not buy a product; they buy what a product does for them. For instance, at a restaurant, a warm personal touch can be more persuasive in winning repeat customers than the food itself. Customer value proposition is what tells customers why they should buy from you and not from the competition (unique differentiation) or that one particular product or service will add more value or better solve a problem than other similar offerings. In the competitive landscape, the company with the best value proposition wins.

There are three strategic weapons with which you can gain competitive advantage:

1. **Time-based** competitive advantage: FedEx, McDonald's, Domino's Pizza, Toyota, Honda, and Sony use agile strategies based on the cycle of flexible manufacturing, trimming process time in the factory, rapid response, and increasing innovation. As Jason Jennings and Laurence Haughton state in the title of their book, *It's Not the Big That Eat the Small—It's the Fast That Eat the Slow.**
2. **Cost-based** competitive advantage: cost-management strategies; cost drivers; low-cost provider strategies.
3. **Quality-based** competitive advantage: delivering defect-free goods and services the first time, and all the time.

Customer satisfaction is a reflection of the state of the business and depends on three elements: delivering a defect-free product or service; delivering a product or service on schedule; and delivering a product or service at the lowest possible cost. In order to have an enduring competitive advantage, it is important to develop a wide spectrum of competitive strategies. The excellent companies try not to focus their competitive strategy on one factor alone but strive always to stay on the cutting edge. They combine multiple competitive strategies to create a lasting

---

* New York: HarperCollins, 2002.

advantage and continue to innovate as they use technology to raise the bar with each method. Yes, you can compete on time–cost–quality, but you can also compete on reputation, values, technology, convenience, service, design, and innovation. You need a potent mix of all of these and more. The essence of sustainable competitive advantage is to achieve differentiation advantage* in order to deliver benefits that exceed those of competing products.

Cultivating such a mind-set could be the secret of your organization's success. You should also take heed of the advice offered by the CEO of Cisco Systems, John Chambers: "Build a culture that thrives on change. After you achieve competitive advantage, others will copy that and commoditize it. Therefore, you need to be ahead of the competition by creating a new competitive advantage faster."

In light of the popularity of customer centricity one might ask, is Apple a great innovator or is it actually tuned into the voice of its customers? Consistent with Apple's marketing motto of "*Think Differently*," Steve Jobs's philosophy of designing products is not "steered by committee or determined by market research." Rather, he "relies heavily on tenacity, patience, belief, and instinct."[†]

As Figure 1.2 illustrates, in the world of supply-chain management the paradigm has shifted from price to cost and then to customer value. Over the years, competitive advantage has shifted from cost differentiation to quality differentiation and now to customer value. Today, it is a given that companies offer competitively priced, quality products. The key to differentiation, therefore, is value to the customer, which entails meeting or exceeding customer needs (in product value and process value).

Businesses in the market economy compete on the basis of quality, price, delivery, and service. A country is like a battleship and is not easy to turnaround; countries compete at a different level. However, a smart government that works better and costs less is a critical ingredient of national competitiveness. **Competitiveness** is the institutional ability of a country to constantly overcome *binding constraints* on growth.

The best-governed state is one in which the citizens are sovereign and the best-managed business enterprise is one in which the customer is king

---

* Michael Porter, *Competitive Strategy: Techniques for Analyzing Industries and Competitors* (New York: Free Press, 1980).

† Steve Lohr, Steve Jobs and the Economics of Elitism, *New York Times*, January 31, 2010, accessed August 31, 2013, http://www.nytimes.com/2010/01/31/weekinreview/31lohr.html?_r=0.

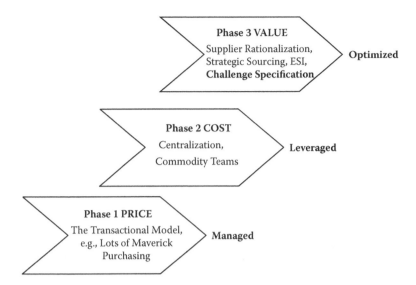

FIGURE 1.2
SCM maturity path.

or queen. In the business world, where there is **competition**, customers enjoy better quality and lower prices. Likewise in government, when there is genuine competition, citizens enjoy better public service because in a competitive environment the government will work better and cost less and thereby deliver best value-for-money.

The concept of competitiveness has two components: *comparative advantage* and *competitive advantage.* Beyond the Ricardian classical theory of comparative advantage, in the current management parlance, the term "comparative advantage" refers to those sectors in which a given country specializes in international trade in view of its resource endowment, which is a national prosperity that is *inherited.* It refers to primary resources and factor endowment in terms of land, labor, and capital. It is a useful *static* approach, which explains, at any given moment, the international division of labor.* (Friedrich von Kirchbach, International Trade Centre/UNCTAD/WTO.)

Competitive advantage is a *dynamic* concept relating to the set of institutions and economic policies supportive of high rates of economic growth in the medium term. It is based on the principle that national prosperity

---

* Presentation by Friedrich von Kirchbach to the International Trade Centre/UNCTAD/WTO.

is created through competitiveness that is driven by the innovation and productivity of private enterprises. National wealth depends more on knowledge than on natural resources. Therefore, investing in people is perhaps the single most important factor in economic growth.[*]

Having said that about the genesis and goals of VA/VE, let us now define the terminology. The simple definition of VA/VE is that it is the management technique that achieves the required function at the lowest cost in resources. Three prominent professional associations specializing in VA/VE, supply management, and operations management have refined their definitions for VA/VE over the years. However, the following summarized three definitions seem to be in common use:

> "**VALUE ENGINEERING** is an organized, systematic, and creative discipline directed at analyzing the function of a product or process with the purpose of achieving the required function at the lowest overall cost consistent with requirements for performance, including reliability, quality, maintainability, safety, and delivery."—SAVE International, the Value Society

> "**VALUE ANALYSIS** is a systematic and objective evaluation of the value of a good or service, focusing on an analysis of function relative to the cost of manufacturing or providing the item or service. Value Analysis provides insight into the inherent worth of the final good or service, possibly altering specification and quality requirements that could reduce costs without impairing functional suitability. Value Engineering is value analysis conducted at the design engineering stage of the product development process."—Institute for Supply Management (ISM)

> "**VALUE ANALYSIS** is a disciplined approach to the elimination of waste from products or processes through an investigative process that focuses on the functions to be performed and whether such functions add value to the products or processes."—American Production & Inventory Control Society (APICS), the Association for Operations Management

---

[*] Michael Porter, The competitive advantage of nations, *Harvard Business Review*, 68(2) (March–April 1990): 185. For a more expanded exposé also see his books on competitive advantage.

In summary, Value Analysis is **a way of thinking**. It is also

- A creative and organized problem-solving method
- A scientific method to achieve operational excellence
- A system of techniques for measuring and controlling value
- A method for identifying and eliminating unnecessary cost
- A proven weapon to attack mediocrity and cheapening
- A creative means of spending money more wisely
- An effective tool to improve the value of a product, process, or project
- A search for better value at lower cost
- A change enabler, driving quantifiable outcomes

2

The Value Equation

WHAT IS VALUE?

"Value" is a difficult word to define because it is applied to both subjective and objective qualities. About the year 350 BC, Aristotle identified seven broad categories of value: economic, aesthetic, political, moral, religious, judicial, and ethical. So, what is new since those early definitions of value? What is new is the term and methodology of VA/VE.

Of the different categories of value listed above, VA/VE focuses on **economic value**. But once you master the value methodology based on economic value, you may find that it can also be applied to social values, like a decision whether to get married or not. It may even be applied to political value, for example, how to replace ethnic politics based on doctrinaire ideology and self-interest of a party with that of a patriotic movement that embraces national interest.

For the purpose of our value study, we will discuss **four kinds of economic value**:

1. **Use Value:** The properties that accomplish a use, work, or service function. The term "use value" encompasses performance, reliability, and quality.
   a. PERFORMANCE is the ability of a product, process, or project to accomplish an intended function.
   b. RELIABILITY is the ability of a product to be acceptably trouble free, to have a satisfactory lifespan. It is the degree of confidence or probability that an item will perform a specified number of times under prescribed conditions.

   c. QUALITY is the percentage of acceptable operating products at a precise moment. It is conformance to the specifications that result in a product or service that meets or exceeds customer requirements or expectations. Six Sigma is a quality model that will be discussed later.

2. **Esteem Value:** The properties or features that make ownership of a product desirable.

3. **Cost Value:** The sum of labor, material, overhead, and all other essential elements of cost that is required to produce an item or provide a service.

4. **Exchange Value:** The monetary measure of properties or qualities of a product that enable it to be exchanged for something else, referred to as PRICE.

THE VALUE EQUATION

From the foregoing definitions of the four kinds of economic value we can formulate a Value Equation (Figure 2.1) of **V equals F over C**; F denoting Function and C denoting Cost, meaning that it is satisfaction of need over resources used. This is the formula for a balanced approach.

We need to be reminded about what happens when a proper balance is not maintained between function and cost. The Wallace Company that won the Malcolm Baldrige Quality Award in 1991 went bankrupt a few years later.

A PRODUCT is anything that is the result of someone's efforts, that is, labor or mental application. It is whatever we give customers for their

FIGURE 2.1
The Value Equation.

money, whether it is a physical product or a service. The word "product" refers to both goods and services. FUNCTION is an element, characteristic, or action that makes the product work or sell. The audacity to challenge specifications is an essential trait of VA/VE. The use of FUNCTIONAL ANALYSIS differentiates VA/VE from other problem-solving approaches.

There are two conditions that are necessary for a product to have economic value: *utility* and *scarcity*.

UTILITY refers to a product's usefulness, its capacity to satisfy a need or serve a purpose. Actually satisfying that need is the function of the product. As management guru Peter Drucker tells us, "Nobody pays for a 'product.' What is paid for is satisfaction."* As you know, the purpose of economic activity is not really to provide commodities, but is fundamentally to provide utility via commodity. Utility is appraised as WORTH. Product worth is an appraisal of the combination of essential and desirable characteristics rendering a product useful or esteemed in the eyes of a buyer; not in the *structural* sense of what the product *is*, but in the *functional* sense of what it *does*. Therefore, every purchase order buys a function. In his classic *Harvard Business Review* article, "Marketing Myopia," Theodore Levitt wrote, "People don't buy ¼" drill bits; they buy ¼ holes."

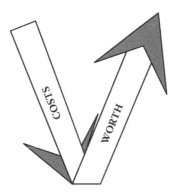

When COSTS go down and WORTH goes up, that's how you get VALUE. Value can therefore be enhanced by either improving the function or reducing the cost; while abiding by the primary tenet of VA/VE that the basic function must be preserved and not reduced, assuming that it is not already overengineered.

PRODUCT VALUE then is the relationship between product worth and product cost. It is the relationship between what performing the function

---

* *Harvard Business Review*, July-August 1960.

is worth to a buyer and what it costs the buyer. Value is the lowest overall cost that must be paid to have an essential function performed reliably. Value is the ratio between a function for customer satisfaction and the cost of that function. Value is anything the customer is willing to pay for. WASTE is anything the customer is not willing to pay for.

SCARCITY refers to a product's difficulty in attainment, and is expressed in monetary terms as cost. The OUTPUTS of organizations are profitable products, and INPUTS are costs that represent resources consumed. PRODUCT COST is everything that the customer has to pay for in order to acquire, use, enjoy, maintain, and dispose of a product. In other words, it includes all the resources that must be utilized to provide whatever the customer wants. All cost is for function. Cost does not exist by itself. It is always incurred, in intent at least, for the sake of a result. No matter how inexpensive or efficient an effort, it is waste, rather than cost, if it is devoid of a result. That will be true also with dysfunctional operations and malfunctioning products. Therefore, WASTE is the cost of efforts that cannot produce results. The purpose of VA/VE is to identify and engineer such unnecessary costs out of products without adversely affecting their functional integrity. The costliest wastes are *not-doing* and *gold plating*. Not-doing is the airplane that is being maintained in a hangar for too long and not making money for an airline. Gold plating is using a five-cent item when a one-cent item can do the job.

We appraise product value as a ratio of **worth to cost**. A product will be good value for me if it costs me $10 but is worth $20. It will be fair value if it costs me $10 and is worth $10. It will be poor value if it costs me $10 but is worth only $5. Obviously, value is determined by comparison only. (Does this question sound familiar? "Are you better off today than you were four years ago?") Value Analysis is your tool for worth/cost analysis, as when you ask, "Is this going to be worth the cost?"

VALUE-FOR-MONEY

Based on the formulation that **Value = Function/Cost** and applying the value methodology of **PISERIA,** your value studies could result in one of the following three types of VA/VE, illustrated in Figure 2.2:

1. **Acceptable VA/VE** will enable you to perform the same function at less cost, resulting in *good value.*

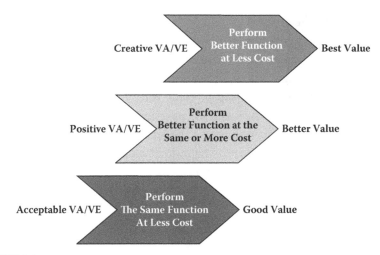

FIGURE 2.2
Demystifying VfM ethos.

2. **Positive VA/VE** will enable you to perform a better function at the same or more cost, resulting in *better value*. (In a growing economy, customers want better products at prices they can afford.)

3. **Creative VA/VE** will enable you to perform a better function at less cost, resulting in *best value*.

Value-for-Money (VfM) is achieved through pursuing the lowest total cost of ownership (TCO), clearly defining relevant benefits, and delivering on time. Preventing waste and fostering competition, transparency, and accountability during the tendering or bidding process are key conditions to achieving VfM. As illustrated in Figure 2.2, while the main objective of VfM is the optimum utilization of resources to achieve the desired results, the outcomes could range from good to better to best value. Therefore, one needs to check for real business impact by performing a VfM audit.

VALUE INNOVATION AND WAYS TO ADD VALUE

Blue Ocean Strategy authors W. Chan Kim and Renée Mauborgne* wrote that **value innovation** occurs only when companies align innovation with **utility** and **cost**. They state that the high-growth companies and market

---

* *Blue Ocean Strategy: How to Create Uncontested Market Space and Make the Competition Irrelevant* (Boston, MA: Harvard Business School, 2005).

leaders who are the creators of blue oceans (uncontested market spaces) pay little attention to matching or beating their rivals. Instead, they follow the strategic logic known as value innovation. In value innovation, instead of focusing on beating the competition you focus on making the competition irrelevant by creating a leap in value for customers and your company, thereby opening up new and uncontested market space. In order to be a perpetually excellent company, you need to consistently make strategic moves for creating and capturing blue oceans. Blue-ocean strategy is about revolutionary value innovation.

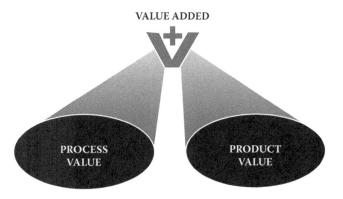

No matter what your position in your company, if you don't add value, you don't matter. When you add value you become a problem solver. Value-added contributions include such measurable roles and activities as saved money, awed customers, increased sales, and improved product (product value), or achieved significant reduction in the time or steps necessary to complete a process (process value). Here are 23 more specific **ways to add process value and product value**:

1. Knock your responsibilities out of the park.
2. Learn something new every week that could impact your company.
3. Implement the new concepts you learned at the last seminar or conference.
4. Go beyond your job description and become known as a problem solver.
5. Use your suppliers' knowledge to better understand opportunities.
6. Document real cost savings and report them to management.
7. Identify what is bought, by whom, and in what amount, and then determine areas for improvement or how you can add value to the process. Who are your customers and what are their requirements?

8. Establish cross-functional teams.
9. Establish internal performance measures.
10. Determine the competency level of your staff and design appropriate training programs to improve their skills.
11. Add value or outsource or dispose of unneeded assets.*
12. Initiate value-improvement projects.
13. Improve time management for yourself and others.
14. Streamline processes by introducing effective systems, such as simplify–substitute–standardize.
15. Reduce unnecessary expenditures by eliminating unauthorized maverick purchasing.
16. Identify and eliminate waste.
17. Increase productivity of staff if you are in a supervisory or management role.
18. Improve employee morale.
19. Ensure that your customer-service skills are second to none and that no customer ever has a justifiable reason to complain about you.
20. Underpromise and overdeliver on your assignments.
21. Become the go-to person who is always on call for solving problems.
22. Demonstrate how you can add value better than your competitors.
23. Learn to sell your contributions and keep your Brag Bag up to date.

---

* When Lee Raymond was CEO of Exxon Mobil, he required the corporate-planning team to identify 3 to 5% of the company's assets for potential disposal every year. Exxon Mobil's divisions were allowed to retain assets placed in this group only if they could demonstrate a tangible and compelling turnaround program. When Jack Welch was CEO at GE, he had a rule to dispose of a business if it was not number 1 or 2 in the industry.

3

Achieving Excellence

U.S. Airways Flight 1549 took off from La Guardia Airport in New York City on January 15, 2009. The plane hit a flock of geese and lost both engines. Quick thinking and his toolkit of knowledge, skills, and experience acquired during his 35-year aviation career allowed Captain Chesley "Sully" Sullenberger to land his aircraft safely on the Hudson River with no loss of life among the 150 passengers and 6 crew members. Sully's performance on that day was a personification of excellence.

Stephen Covey, author of *The 7 Habits of Highly Effective People*,* advises us to "begin with the end in mind." With VA/VE, we are engaged in the pursuit of excellence, thus making VA/VE the driver of operational excellence. **Operational Excellence** is a comprehensive approach to leading more efficient and effective organizations by integrating the continuous value-improvement methods mentioned under the introduction, thereby making the whole greater than the sum of its parts. In its simplest terms, operational excellence means consistently doing things well across the value chain as a way of gaining competitive advantage. In its broadest terms, it is a discipline that drives corporate strategy.

Jim Collins, author of *Good to Great: Why Some Companies Make the Leap... And Others Don't*, believes that "in the end, it is impossible to have a great life unless it is a meaningful life. And it is very difficult to have a meaningful life without meaningful work."† The practice of VA/VE makes work meaningful by providing people with the opportunity for self-actualization and achievement, thereby helping them make a difference.

---

* New York: Free Press, 2004.
† New York: Harper Business, 2001.

PRINCIPLES OF EXCELLENCE

If you want your employees to succeed by excelling in what they do and help them transform your organization into a center of excellence, you need to nurture a mind-set that embraces the following **principles of excellence**:

- Focus on improving customer satisfaction by reducing variation, cycle time, defects, and eliminating waste. Enhance value to the customer.
- Listen to the voice of the customer and use VoC data/information/knowledge to drive operational excellence.
- Establish stretch goals or aggressive rate-of-improvement goals to increase to fivefold, then to tenfold your resource productivity through better VA/VE.
- Generate sustained success by creating an organizational culture of achievement and constant renewal; improving and innovation, non-stop; forever evolving, never standing still!
- Strive for excellence—achieve six-sigma quality at lean speed. Build a better mousetrap.
- Popularize continuous value improvement through "learn-and-apply" value project initiatives.
- Develop an entrepreneurial-minded workforce through value training.
- Nurture a mind-set driven by best *value-for-money* ethos.

With a mind-set that has excellence as the overarching theme and the use of the research methodology of VA/VE, you will be able to improve quality, compress time, and slash cost, that is, optimize resources by adopting the new mantra of doing work **better, faster, and leaner** than the competition. Can you imagine being given a work assignment and being told, "Take as much time as you want, spend as much money as you need, and I don't care how well it is done?" And if you were on the operating table in a hospital, how much error would you allow your surgeon? We demand excellence in our daily lives and need to make it our constant business goal as well.

VA/VE is not typical cost reduction because it doesn't cheapen the product or service. Cheapening means reducing cost at the expense of performance. In the VA/VE world of lean thinking, the word "cheaper" is forbidden as it implies poor quality or mediocrity; and, as Benjamin Franklin said, "the bitterness of poor quality remains long after the sweetness of low price is forgotten." For instance, consider the application of value-mindedness in a hospital environment where a patient expresses delight with the medical team for solving the problem effectively, without keeping the patient waiting too long, and at an affordable cost! In this regard, it is worth taking note of the current Cost, Quality, and Outcomes (CQO) Movement that is being pursued as a best practice by health care supply-chain professionals, where application of value analysis is baked into the operating principles of daily life at work. Medical-center administrators score their physicians according to data in efficiency (length of stay, cost per case), effectiveness (readmission rates, mortality rates), and patient perception of care (patient-satisfaction survey scores). Currently, health care is facing a new era of rapid transformation that will focus the industry on payment for value rather than transactions.

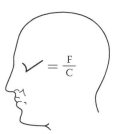

Value-minded people think and work differently. We are in the business of doing work (F for "function") at less cost (C). If we can't do everything better, faster, and leaner than our competitor we'll be out of business before we know it; then we won't have a mission. Albert Einstein defined insanity as doing the same thing over and over again and expecting different results. In business this means do it differently from the way you are used to doing things if you want different results. Value hunters think differently as their goal is to assure best value for money.

THE EXCELLENCE MODEL

Excellent companies lead in the areas depicted in Figure 3.1. How does your company measure up? Which strategy is your forte? What is the secret of your success—customer intimacy, product leadership, operational

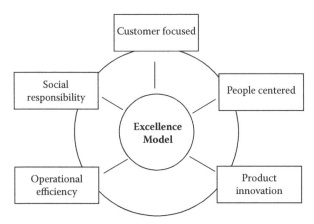

FIGURE 3.1
Excellence model. (Adapted from Hagel, J. III and Singer, M., *Harv. Bus. Rev.*, 133, 1999.)

excellence, or taking good care of the human side of your enterprise? Does your organization create an environment for continuous improvement and purpose maximization by helping employees develop the intrinsic motivation they need to participate in workplace activities such as change management? The richest experiences in our lives are when we are doing something that matters, doing it well, and doing it in the service of a cause larger than ourselves.

4

The Integrated Value-Improvement Methodologies

The integrated PISERIA value methodology has the potential to unify fragmented business processes by disrupting the focus from the single-silo approach of separate, hierarchical functions to a single integrated architecture that optimizes the horizontal flow of products and services through entire value streams, using interconnected teams and networks across various functions to fulfill customer requirements. Such optimization through the PISERIA value methodology will need multidisciplinary, whole-system thinking, thereby creating an ecosystem of improvement methodologies that are linked together through PISERIA. To illustrate, a car can perform its desired function only as the integrated system it was designed to be. Likewise, only a whole system, all of whose elements are in place and mutually coordinated with one another, can enable a company to be an agile competitor.

There are many books published on the change management tools of Lean Thinking, Six Sigma, Quality Function Deployment, and Project Management. My intent here is to show the relationship of these approaches to VA/VE. Briefly, **Lean Thinking** focuses on reducing waste, cycle time, and complexity. **Six Sigma** focuses on reducing variation, defects, and errors. **Quality Function Deployment** transforms customer requirements into design quality. **Project Management** focuses on initiating value improvement through project teams. In addition, **Business Process Reengineering** (BPR), **Kaizen**, and **Total Quality Management** (TQM), discussed in Chapter 5, are integrated into the value methodology.

These tools need to be applied in the right context with the value methodology's integrative problem-solving discipline, to improve productivity, quality, cost, delivery, and service. The PISERIA value methodology transforms the disciplinary silos (Six Sigma, Lean Thinking, and so on, characterized by vertical thinking) into an integrated whole system. As Craig Charles put it "It's *evolve or die*, really, you have to evolve, you have to move on otherwise it just becomes stagnant." Since they are extensively discussed in various publications, I provide here only a summary of these approaches.

LEAN THINKING

According to James Womack of the Lean Enterprise Institute, the core idea of lean thinking is to maximize customer satisfaction while minimizing waste. Lean production (doing more with less) is not about eliminating people, but about expanding capacity by reducing costs and shortening cycle times between order and ship date. With the Toyota Production System, best known for its lean thinking, lean production focuses on eliminating waste in processes (i.e., the waste of work in progress and finished goods inventories) and equipment stops running when it detects a defect.

Listed here are 14 fundamental **Lean Principles** to help you increase your awareness of **waste**. I refer to these as the Public Enemy List as they represent **Non-Value-A**dded activities. NVA is work that the customer is not willing to pay for. Value-add time is any process step or activity that transforms the form, fit, and function (FFF) of the product or service for which the customer is willing to pay. The Japanese word for waste is *muda*.

1. Waste of Defects
2. Waste of Inspection
3. Waste of Unnecessary Motion/Energy/Transaction
4. Waste of Excess Inventory and Bloated Bureaucracy
5. Waste of Not Meeting Customer Requirements
6. Wasteful Spending (Waste, Fraud, and Abuse)
7. Waste in Overengineering
8. Waste of Waiting, Delay, Downtime, and Handoffs
9. Waste of Poor Communication
10. Waste of Unnecessary Transportation and Travel
11. Waste of Rejects/Reworks
12. Not getting it right the first time, all the time
13. Waste of Overproduction
14. Waste of Errors

As a case in point, a new report from the influential Institute of Medicine (IOM) has concluded that the U.S. health-care system is hopelessly broken, blowing $750 billion a year—or about 30 cents of every dollar that goes into it—on unnecessary procedures, paperwork, fraud, price gouging, and other inefficiencies. The social cost of the present profit-maximizing insurance system is enormous and negatively impacts the national budget and security of the United States. The moral and economic arguments are overwhelming, with one of the richest nations in the world having more than 48 million of its citizens uninsured as of 2011 and its global competitive advantage hampered by the unnecessary cost driver of the current health care system. In VA/VE we go beyond cutting costs; we do a more surgical approach for improving the health of health care.

If you want to achieve operational excellence, you need to **Get Rid of Waste** by launching VA/VE initiatives. The various elements of lean thinking are integrated into the value methodology (PISERIA) and addressed at length in Section II of the book.

SIX SIGMA

Six Sigma* is a term coined by Motorola to express process capability in parts per million. The late Bill Smith of Motorola is credited with developing the Six Sigma measurement system under the leadership of CEO Bob Galvin. GE adopted Six Sigma under the leadership of CEO Jack Welch.

Six Sigma is a systematic, disciplined, and data-based method to get to root causes to solve problems and analyze and improve processes with the ultimate goal of achieving no more than 3.4 defects per million opportunities.[†]

Six Sigma is a statistical measure of how well you are meeting customer requirements; it is a measure of the performance of a process or a product in terms of defects such as medical mistakes, lost luggage on flights, or inefficient loan processing time. "Sigma" is a term used in statistics to represent standard deviation, an indicator of the degree of variation of output in a set of measurements or a standard. In its business use, it indicates

---

[*] Greg Brue, *Six Sigma for Managers* (New York: McGraw-Hill, 2002), pp. 2–8; Peter S. Pande, Robert P. Neuman, and Ronald R. Cavanagh, *The Six Sigma Way* (New York: McGraw-Hill, 2000), pp. 19–28.

[†] Bill Carreira and Bill Trudell, *Lean Six Sigma That Works: A Powerful Action Plan for Dramatically Improving Quality, Increasing Speed, and Reducing Waste* (New York: AMACOM, 2006), p. 157.

defects in the output of a process, and helps us understand how far the process deviates from perfection.

The Six Sigma process begins by identifying the critical-to-quality (CTQ) elements or issues of a process. The next step is to count the number of defects that occur, and then calculate the "yield" of the process (percentage of items without defects). Finally, a table is used to determine the Sigma level. Six Sigma is a statistical measure of process capability and performance. Achieving Six Sigma means that your processes are delivering only 3.4 defects per million opportunities (DPMO), which is nearly perfect (99.99966 or 99.9997% yields—just .0003% short of zero defects). Six Sigma is a disciplined extension of Total Quality Management (TQM).

QUALITY FUNCTION DEPLOYMENT

Developed by Dr. Yoji Akao in Japan in 1966, Quality Function Deployment (QFD)* embeds the voice of the customer (VoC) into the design process, ensuring consistent deployment of customer requirements. The *Gemba* (a Japanese term meaning "real place") is a visit to a company during which these customer needs are evidenced and compiled in order to create value for the customer. While QFD uses market research, focus groups, interviews, and surveys to collect VoC, Gemba visits are not scripted or bound by what one wants to ask.

Converting customers' needs into specific product or service design features or functional requirements establishes the formal integration of the QFD processes and value engineering in describing customer expectations in terms of function.

PROJECT MANAGEMENT

The integration of project management into the value methodology of PISERIA is self-evident as all value improvements take place project by project. When an organization observes a problem, it tasks a team with performing a value study or assigns a team to initiate a project. Therefore, the value study is referred to as a "project."

---

* Yoji Akao, *Development History of Quality Function Deployment, The Customer Driven Approach to Quality Planning and Deployment* (Tokyo Japan: Asian Productivity Organization, 1994), p. 339.

Project work is differentiated from regular work, the latter often referred to as "operational work." Operations are ongoing and repetitive, while projects are temporary and unique. However, both are performed by people; constrained by limited resources; and they are planned, executed, and controlled.

A project is a one-time multitask job that has definite starting and ending points; a well-defined scope of work; a budget, and a temporary team that will be disbanded once the job is completed.

Here are different perspectives on the definition of projects and project management:

> "A project is a multitask job that has performance, cost, and time requirements that is done only one time."—James P. Lewis
> "A project is a problem scheduled for solution."—Dr. Joseph M. Juran
> "For the purpose of Value Studies, a project is the subject of the study."— SAVE International
> "Project management is the application of knowledge, skills, and techniques to project activities to meet the project requirements."— PMBOK Guide, Project Management Institute

SUPPLY CHAIN MANAGEMENT AND THE SUPPLY VALUE CHAIN

Supply Value Chain

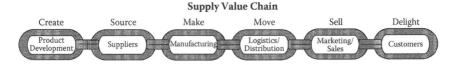

A firm creates value by performing a series of activities that Michael Porter identified as the value chain, and we understand a chain to be only as strong as its weakest link.

Supply chain enterprise is a dynamic supply-and-demand network. Supply chain management integrates supply and demand within and across companies. Traditionally, organizations are vertically structured and view themselves as a collection of independent functions that convert inputs into products and services for customers. Every function, separately and collectively, is expected to focus on the customer and deliver added value. Such vertically structured organizations are measured on the performance of each small piece of the fulfillment stream. But such a setup supports only isolated functional silos and leads to continually optimizing a piece of the stream at the expense of the whole. Silo thinking or verticalization suboptimizes. Organizations should reorient themselves away from managing functions toward facilitating interconnected horizontal processes in the supply value chain. With horizontal management, organizations will be able to work in teams and networks across organizational silos, and that would be central to good governance.

The question is, how much of each department's cost is really value added and how much is merely waste added?

The fundamental systems-interactive paradigm of organizational analysis features continual stages of input, throughput (processing), and output, which demonstrate the concept of openness and closedness. A closed system does not interact with its environment. It does not take in information and therefore is likely to atrophy, that is to vanish. An open system receives information, which it uses to interact dynamically with its environment. Openness increases its likelihood to survive and prosper.

Because organizations operate as functional silos, no one person owns the entire process. The tool for creating a high-level map of the supply value process is known as **SIPOC** (Suppliers, Input, Process, Output, Customers). The SIPOC Diagram (Figure 4.1) is a systems model, a process flow perspective. It is a business process model represented by the

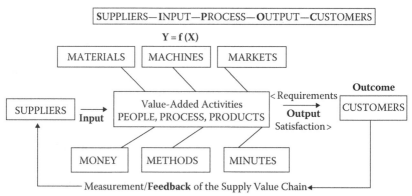

The lifeblood of the corporate body flows through its **resource allocation** system.
WASTE is the use of any resource beyond the minimum amount that is absolutely essential to add value to a product or process.

FIGURE 4.1
SIPOC diagram. (Adapted from Pande, P. et al., *The Six Sigma Way*, McGraw Hill, New York, 2000, pp. 168–170.)

mathematical formula **Y = f (X).** The formula shows that the OUTPUTS are dependent on PROCESS performance (transformation) and quality of the INPUTS.

Y stands for the measures of the end results of the process, that is, outputs and outcomes. OUTPUT measures focus on immediate results such as deliveries, defects, and complaints. OUTCOME measures focus on longer-term impacts such as profit, customer satisfaction, and meeting objectives like cycle-time reduction.

X stands for the measures of INPUTS—the resources applied.

F stands for the PROCESS that converts inputs into outputs. Process is the combination of people, equipment, materials, methods, and so on that produce output (a given product or service).

The Institute for Supply Management (ISM) defines supply chain management (SCM) as a systems approach to integrating the entire flow of information, materials, and services from raw-materials suppliers through factories and warehouses to the end customer. The PISERIA value management methodology further integrates the **SCM's Tool Kit** that consists of the following (and more): contract negotiation, competitive bidding, hedging, standardization, systems contracting, supplier-managed inventory, consignment buying, global sourcing, group purchasing, e-procurement, procurement card, outsourcing, and countertrade. As illustrated in Figure 4.2, operational excellence is achieved by integrating these performance-enhancing

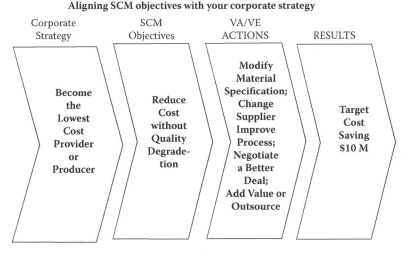

FIGURE 4.2
VA/VE: The enabler of operational excellence!

tools to effectively solve SCM problems. The problem of having only one tool is best explained by Abraham Maslow, who wrote, "I suppose it is tempting, if the only tool you have is a hammer, to treat everything as if it were a nail."[*]

The traditional main source of information about business performance has been an accounting system that is incapable of tracking loss of business, wasted time, inefficiencies of business processes, and other hidden costs. But in today's complex organizations, operational performance can be better improved as exemplified in Figure 4.2 through the corporate strategy of VA/VE as the enabler of operational excellence.

To summarize, the optimization that can be achieved from such integrated improvement methodologies that are linked together through PISERIA is like the different instruments in an orchestra that play music in harmony to produce a delightful melody.

---

[*] Abraham H. Maslow, *The Psychology of Science: A Reconnaissance* (New York: Joanna Cotler Books, 1966), p. 15.

5

Change Management

As you are aware, the only constant in life is change and it is a constantly moving target. If change is an inevitable reality of life then we need to learn how to cope with it. According to Carlos Fallon, there are three **tactics for coping with change**:

"Ignore change and let it happen to you,
Anticipate change and adapt to it, or
Create change and benefit from it."*

I leave it to your imagination to think of the myriad examples of change in the economic and political arenas for each of these three categories. While you are at it, keep in mind that adaptability is the new competitive advantage and that it will be the next wave of management theory. A popular quote that is misattributed to Charles Darwin postulates the same: "It is not the strongest of the species that survives, nor the most intelligent, but rather the one most adaptable to change."

Change management is an approach of transitioning individuals, teams, and organizations from a current state to a desired future state. While there are plenty of change-management models advocated by various consultants and academics, the value-improvement change-management model that I present here follows a learn-and-apply three-phase approach:

1. Value Ability: Preparing and enabling the organization with knowledge and skills of the seven-step process of the PISERIA value methodology that integrates VA/VE with other change-management tools, is to be discussed in detail in Section II.
2. Value Climate: Creating a climate that is conducive to change.

---

* *Value Analysis to Improve Productivity* (Hoboken, NJ: John Wiley & Sons, 1971), p. 178.

3. Value Experience: Implementing and sustaining change through blending classroom training with application to jobs in order to provide contextual learning, thereby achieving results-driven organic rather than mechanistic change.

THE CHANGE EQUATION

A simple, yet most powerful change-management or organization-development model that you will find useful is the Change Equation (also referred to as the Formula for Change) developed by Richard Beckhard and David Gleicher, and modified here to include additional factors. The original formula was created by David Gleicher and published by Beckhard and Harris as $C = (ABD)>X$, where C stands for change, A is status quo dissatisfaction, B is a desired clear vision, D is practical steps to the desired state, and X is the cost of the change. The same formula is also expressed as $D \times V \times F > R$, which stands for **D**issatisfaction with the status quo, **V**ision of what is possible, and **F**irst steps in the direction of the vision. **R**esistance to change will occur if these factors are missing. It is said that "the only one who truly likes change is a wet baby." That's because the baby has a constructive discontent with the current state of affairs.

As provided in the modified version (Figure 5.1), in business, if factors A, B, C, D, and E are not present, then you can expect resistance to change. When the five factors of embracing change are greater than the resistance

The Change Equation

Change = F (A, B, C, D, E) > Resistance

A = Sufficient dissatisfaction with the current state
and belief that change is possible
B = Clear and compelling vision of the future
C = Awareness of the need for change
D = A scientific method to effect the change
E = The benefit will exceed the cost of change

FIGURE 5.1

The change equation. (Adopted from *ChangeEquation* by Beckhard, R. and Gleicher, D., *Organization Development: Strategies and Models*, Addison-Wesley, Reading, MA, 1969.)

to change, then change can happen. People embrace change when they believe the value of change is worth the pain. Change can also happen when the pain of the current state exceeds the anticipated pain of change.

Leaders or employees who persistently resist change create measurable impacts on an organization, including decreased productivity, customer dissatisfaction, loss of valued employees, and the possibility of total failure of a planned change. One of the best ways to lessen resistance to change is to allow employees to share in the decision-making process. This is not always possible, but change works best when employees have had a part in planning it. People like change that is imposed on them the least. People welcome change if they believe that they are the initiators rather than mere recipients of the change. Resistance is diminished when everyone shares in the planning process and understands the benefits of change. Giving people a certain degree of control over their work fulfills the need for freedom and provides opportunity for taking joy in work.

Historically, the change equation can be seen as a major milestone in the field of organization development. The underlying philosophy of Frederick Taylor's scientific management approach was that "workers work, managers think." Taylor's method was a reflection of his times. Today's employers understand the connection between employee involvement and organizational success.

Whatever the situation is, when change looms on the horizon, chances are that you'll hear reactions like, "If God meant us to fly, he would have given us wings." Figure 5.2 shows samples of negative responses to value-improvement suggestions.

Negative Thinking—Closed Mind

- That is not my job.
- It won't work in our territory.
- It is not practical.
- We tried that before.
- It is against company policy.
- It is too radical a change.
- We don't have the time for that.
- Don't be ridiculous.
- We are too small a territory.
- We have never done it before.
- It will be too hard to sell.
- The union will scream.
- You are two years ahead of your time.
- Let us get back to reality.
- It isn't in the budget.
- That is not our problem.
- Why change it; it is still working ok.
- We are not ready for that.
- You can't teach an old dog new tricks.
- The top brass would never go for it.
- We did alright without it.
- Let us shelve it for the time being.
- It is just the flavor the month
- Too busy to study it.
- Has anyone else tried it?
- Let us form a committee.

FIGURE 5.2
Negative thinking.

Positive Thinking—Open Mind

- It looks promising, show me.
- Let us think about this.
- How does it compare?
- How does it work?
- How can we find out more about it?
- Let us test it.
- Let us find out.
- Let us try this in one spot, anyway.
- Where might it work best?

- What would be a fair test period?
- What is the best way to get started?
- How can we improve on it?
- Who else might have some ideas on it?
- What items should we start with?
- Let us see how it looks.
- How much time do we have?
- What do you think?
- When do we start?
- If it ain't broke, make it better.

The mind is like a parachute...it works only when open.

FIGURE 5.3
Positive thinking.

On the other hand, if you wish to mobilize commitment and reduce barriers to change, you would want to promote employee motivation. Understanding the human side of change and mastering the "soft" side of change management is essential for organizational success. Companies will reap the rewards only when change occurs at the level of the individual employee. Leadership teams that fail to plan for the human side of change often find themselves wondering why their best-laid plans have gone awry. By contrast, Figure 5.3 presents those positive-feedback statements that could help the move toward the realization of continuous value improvement.

THE PROCESS OF CHANGE

The following brief is adapted from *The Handbook of Conflict Resolution: Theory and Practice.**

Two authors best known for theorizing about the process of change are Kurt Lewin and Richard Berkhard. Lewin provided an overall theoretical framework for understanding the process of change, which he identified as **unfreezing, movement,** and **refreezing**. Lewin's concept is a linear

---

* Edited by Morton Deutsch, Peter T. Coleman, and Eric C. Marcus (New York: Jossey-Bass, 2006), pp. 437–433.

description that is often applied to understanding change in both individuals and social systems.

Lewin applied **force-field analysis** to understand the unfreezing process. Force- field analysis illustrates the current state of the system. A force field shows the relationship between factors and forces that help promote a change and those that oppose or create resistance to change. Lewin refers to forces that promote the change goal as *driving forces* and those that work in opposition to it as *restraining forces*. The positives can be reinforced and the negatives eliminated or reduced. To begin the process of change or unfreezing from the current state, the driving forces must be relatively stronger than the resisting forces.

The **transition or movement** takes some action that changes or moves the social system to a new level. Restraining forces, which are also a form of resistance to change, will make the transition difficult. The degree of resistance has an impact on the ease of unfreezing. Here is where *change acceleration process* becomes critical.

Refreezing involves establishing actions or processes that support the new level of behavior and lead to resilience against those resistant forces encouraging old patterns and behaviors. Refreezing may also be understood in terms of the degree of commitment to the new, changed state that exists in the system.

Beckhard applied Lewin's concepts as **Current State**, **Transition State**, and **Desired Future State**. He applied Lewin's concept to understand planned change. Through linear looking, Beckhard suggests that beginning with the end helps to establish a goal for the change and serves the purpose of beginning the process of unfreezing. Furthermore, starting with what people desire in the future generates enthusiasm, motivation, and commitment to the plan and to its implementation. Beckhard's model brings out the conflict inherent in the process of unfreezing by identifying the gaps between the current state and the desired future state.

VA/VE can be used as the method to provide feedback to the system. Feedback or information obtained about the current system by applying the value methodology will help to understand the gaps between the current state and the desired future state and thereby increase awareness of the need for change.

Figure 5.4 is a checklist to help initiate a change process, referred to as "The Drivers of Change" or critical success factors: (1) Eliminate, (2) Streamline, (3) Automate, and (4) Professionalize. Those drivers must be followed in that order, as you would not be expected to automate what must be eliminated.

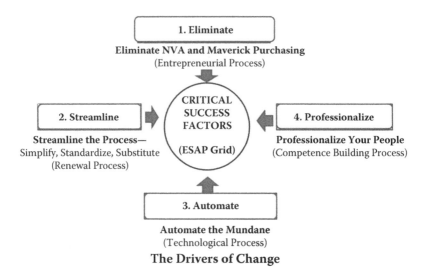

FIGURE 5.4

The drivers of change.

THE FORCES OF CHANGE

The two forces of change (Figure 5.5), also known as value-improvement strategies are known in Japan as **Kaikaku** and **Kaizen** while in the United States they are known as **Business Process Reengineering** (BPR) and **Total Quality Management** (TQM) or Continuous Value Improvement.

Michael Hammer and James Champy,* refer to BPR as the radical redesign of a company's processes, organization, and culture. They state that the aim of reengineering is a quantum leap in performance—the 100% or even tenfold improvement that can follow from entirely new work processes and structures.

BPR is about fundamentally rethinking and radically redesigning business processes in order to achieve dramatic improvements in quality, cost, speed, and service. Here are samples of the disruptive changes of such revolutionary efforts. Major transformations took place in the computer world, from vacuum tubes, to transistors, to microchips. Other such dramatic events include changes from bank tellers to ATM machines;

---

* *Reengineering the Corporation: A Manifesto for Business Revolution* (New York: Harper Business Essentials, 1993).

Mastering Change

Item \ Forces of Change	BUSINEES PROCESS REENGINEERING *Kaikaku*		CONTINUOUS VALUE IMPROVEMENT/TQM *Kaizen*
PROCESS:	Wrong		Basically Right
CHANGE:	Transform/Revolution		Reform/Evolution
SPARK:	Breakthrough/Replace		Enhancement/Fix
PACE:	Rapid/great-leap-fwd		Gradual progress
SCALE:	Dramatic/Step Change		Incremental
LEVEL:	Macro/Sponsor Driven		Micro/Work Unit Driven
INVESTMENT:	High (technology & $)		Low (people)
ORIENTATION:	Customer Value-Added		Customer Value-Added
OUTCOME (GOAL):	Increase quality, cut costs, reduce cycle time, reduce defects, eliminate waste, improve customer satisfaction, empower teams.		

> **VA/VE METHODOLOGY: The whole-system thinking of PISERIA integrates the principles of VA/VE with Six Sigma, Lean Thinking, PM, BPR, and TQM.**

The essence of sustainable competitive advantage revolves around an organization's ability to continually differentiate itself in the marketplace through breakthrough possibilities and frequent enhancements.

FIGURE 5.5

The forces of change.

from classroom teaching to virtual learning; from typewriters to computers; from incandescent lamps to fluorescent to LED, just to mention a few.

As depicted in Figure 5.5, BPR is about episodic, transformational, step-changes, while TQM is about continuous change that is evolving and incremental. The BPR question is, "Are we doing the right things?" The TQM question is, "Are we doing things right?" Therefore, *the framework for mastering change* is enhancing an organization's ability to continually differentiate itself through breakthrough possibilities and frequent enhancements by applying both BPR and TQM.

Both radical innovation and incremental improvement are essential approaches to change management. However, it appears that BPR is more prevalent in the West, meaning that the West seems to prefer more significant improvements made less often while the Japanese favor smaller increments of improvement but introduce them more often. Many believe that Kaizen only allows small improvements to occur and that a great leap forward is not possible. But the cumulative effect of many small steps can be dramatic improvement from a strategic perspective.

Regardless of whether you are in the West or the East, TQM or continuous improvement is not always the right approach. In some cases the reengineering approach may be necessary but it will depend on the circumstances because management is situational. Both BPR and TQM add value; they are complementary and one does not replace the other. Hence, organizations need to know how and when to use both.

Kaizen is made up of two Japanese words, *Kai*, meaning "continuous" (literally "change" or "to correct") and *Zen*, meaning "improvement" (literally "good" or "for the better"). Therefore Kaizen means gradual, unending improvement, doing things better continuously.

Both Kaizen and TQM deal with the topic of quality. TQM is a philosophy of what makes up a quality organization, and it is a prime component of Six Sigma. Kaizen is a methodology for improving existing processes. Kaizen was created in Japan following World War II. It is believed to be the simple truth behind Japan's economic miracle. W. Edwards Deming is generally recognized as the philosopher-guru of TQM. Japan embraced Deming's quality ideas enthusiastically and even named their premier annual prize for manufacturing excellence after him, as The Deming Prize in 1951.

Deming developed the set of fourteen management principles listed in Figure 5.6. Deming's principles 2 through 5 would imply the need for some sort of quality improvement and continuous process-improvement methodology such as Kaizen. So this would make Kaizen a subset of TQM and not a peer concept for comparative purposes.

Many countries have established national quality awards or business excellence awards to recognize deserving organizations, such as the Malcolm Baldrige National Quality Award in the United States, the UK Business Excellence Award, Singapore Quality Award, Mauritian National Quality Award, China Quality Award, and India's Najiv Ghandi National Quality Award, to name a few. For more examples, visit the Centre for Organizational Excellence Research website, www.coer.org.nz.

Deming's 14 Points

Deming, who was credited with jump-starting Japan's economy to its heights, identified 14 organizational transformation points.

1. Create constancy of purpose for improving products & services
2. Adopt philosophy of prevention
3. Cease mass inspection
4. Select a few suppliers based on quality, not on price tag alone
5. Constantly improve system of work
6. Institute modern methods of training on the job
7. Institute leadership development
8. Eliminate fear among employess
9. Eliminate barriers between and among departments
10. Eliminate slogans
11. Remove numerical quotas
12. Enhance worker pride
13. Institute vigorous training and educational programs
14. Implement these 13 points

FIGURE 5.6

Deming's 14 points. (Adapted from Deming, W.E., *Out of the Crisis*, MIT, Cambridge, MA, 1982.)

Section II

The Value Methodology

In fact, it is to children that the scientific method should be taught, for it must be instilled early. If a child grows up without this mental discipline and becomes an adult without having learned how to think in a systematic way, it may be too late to begin then.

—**Isaac Asimov,** in his introduction to: *Science Fare: An Illustrated Guide and Catalog of Toys, Books, and Activities for Kids,* by Wendy Saul and Alan R. Newman

6

Overview of Thinking Methodologies

In order to appreciate the power and relevance of the scientific method of the Value Methodology, let us first review the different problem-solving approaches in practice.

INTUITIVE THINKING VERSUS VALUE THINKING

Daniel Kahneman, recipient of the Nobel Prize in Economic Sciences and author of *Thinking, Fast and Slow*,* describes two systems that drive the way we think: "the intuitive **system 1**, which does the automatic and fast thinking, and the effortful **system 2**, which does the slow thinking." He further elaborates by stating,

> Jumping to conclusions on the basis of limited evidence is so important to the understanding of intuitive thinking, that I will use a cumbersome abbreviation for it: WYSIATI, which stands for 'what you see is all there is.' System 2 is capable of a more systematic and careful approach to evidence and of following a list of boxes that must be checked before making a decision, such as thinking of buying a home, when you deliberately seek information that you don't have.[†]

Author Malcolm Gladwell reminds us in his book *Blink* (p. 143), "When we talk about analytic versus intuitive decision making, neither is good or bad. What is bad is if you use either of them in an inappropriate circumstance. If you get too caught up in the production of information, you drown in the data."

---

* New York: Farrar, Straus, and Giroux, 2011, p. 408.
[†] p. 86.

INDUCTIVE REASONING & DEDUCTIVE REASONING

Former McKinsey consultants Ethan Rasiel and Paul Friga elucidate inductive and deductive reasoning in their book, *The McKinsey Mind*[*]:

> Having your conclusions or recommendations up front is sometimes known as **inductive reasoning**. This approach starts with an initial hypothesis based on intuition. Inductive reasoning takes the form of "We believe X because of reasons A, B, and C." This contrasts with **deductive reasoning**, which can run along the lines of, "A is true, B is true, and C is true; therefore, we believe X." It is obvious then by starting with your conclusion (inductive reasoning), you prevent your audience from asking, "Where is she going with this?" Whereas it will be very easy to lose your audience before you get to your conclusions in data-intensive presentations (deductive reasoning).

Inductive reasoning is also called *generalizing* as it takes specific instances and creates a general rule. When data doesn't exist, organizations make decisions based on hunches and intuitions. Deductive reasoning is data-driven thinking. But data itself isn't the solution; it is just part of the path to the solution. While inductive reasoning goes from a set of specific observations to general conclusions, deductive reasoning flows from general to specific. From general premises, a scientist would extrapolate to specific results. This is a prediction about a specific case based on the general premises. This type of detailed problem and risk analysis helps you to make an unbiased decision. By skipping this analysis and relying on gut instinct, your evaluation will be influenced by your preconceived beliefs and prior experience. One of the shortcomings of intuition is that it could suffer from theory-induced blindness. But to be an effective problem solver, you need to be systematic and logical in your approach. Don't simply trust intuitive judgment. But don't dismiss it, either, as sometimes it would be possible for you to make good snap judgments under conditions of stress.

DMAIC

We now turn to the change-management tools like Six Sigma and Lean Thinking (discussed in Chapter 4) and review their problem-solving methodologies. Organizations that have adopted Lean Six Sigma use

---

[*] *The McKinsey Mind: Understanding and Implementing the Problem-Solving Tools and Management Techniques of the World's Top Strategic Consulting Firm* (New York: McGraw Hill, 2001), p. 111.

the **DMAIC** methodology to identify and eliminate the causes of defects. DMAIC stands for Define, Measure, Analyze, Improve, and Control.

While the principles of Lean Six Sigma are well appreciated by many, there are some critics of its DMAIC methodology. Keki R. Bhote, who was instrumental in launching the Six Sigma process at Motorola, wrote the following commentary about DMAIC: "DMAIC is nothing but a warmed-over version of PDCA (plan, do, check, act). DMAIC is muddled in **D**efinition, imprecise in **M**easurement, impotent in **A**nalysis, incapacitated in **I**mprovement, and rudderless in **C**ontrol."[*]

Bhote is also critical of Six Sigma's emphasis on the elitist organization of black belts and master black belts and prefers the "everybody a black belt" principle.[†] In other words, he advocates a practice of something everyone does rather than an elitist approach. In a similar vein, Raytheon also does not follow the DMAIC model and has developed its own six steps for Six Sigma: Visualize, Commit, Prioritize, Characterize, Improve, and Achieve.[‡]

The value methodology (PISERIA), which is based on the scientific method, is offered in this book as an alternative to DMAIC as it could also serve as the standard operating procedure for all the well-known improvement methodologies like Six Sigma, Lean Thinking, Project Management, BPR, Kaizen, TQM, VA/VE, and others.

PROJECT MANAGEMENT PROCESS

Since all improvements take place project by project, and because project management is an outgrowth of systems management, we need to visit and appreciate the relationship between the **project-management process** and value studies that are VA/VE projects. The Project Management Institute (PMI) defines project management as "the application of knowledge, skills, tools and techniques to a broad

---

[*] *The Power of Ultimate Six Sigma: Keki Bhote's Proven System for Moving Beyond Quality Excellence to Total Business Excellence* (New York: AMACOM, 2003), p. 13.

[†] Ibid., pp. 18, 24.

[‡] Bob Blair and Job McKenzie, Raytheon: New Challenges, New Solutions, and Documented Results. *Defense Acquisition Review Journal*, August 2004, http://www.dau.mil/pubscats/PubsCats/AR%20Journal/arq2004/Blair.pdf.

range of activities in order to meet the requirements of a particular project." According to PMI's *Guide to the Project Management Body of Knowledge* (PMBOK Guide),* project-management *knowledge* draws on the following 10 areas:

- Project integration management
- Project scope management
- Project time management
- Project cost management
- Project quality management
- Project human resource management
- Project communications management
- Project risk management
- Project procurement management
- Project stakeholders management (added in the 5th edition)

Projects require a process just as in problem solving and the phases of a project closely resemble the stages of problem solving. The project-management process (or steps in managing a project) consists of Initiating, Planning, Executing, Monitoring and Controlling, and Closing. Information about each step of the project management process is provided as follows:

1. Project initiation: Needs/requirements; scope; spend analysis; stakeholder analysis; project charter, deliverables.
2. Project planning: What must be done? Who will do it? How will it be done? When must it be done: How much will it cost? What do we need to do it? Scope, roles, and responsibilities, responsibility assignment matrix (RAM), budget and schedule, risk and issues management (proactive, reactive), watch for scope creep.
3. Project execution: Project team management; quality assurance of deliverables; conduct procurement.
4. Project monitoring and control: Constraints (scope, time, cost), earned value (track costs to be on time and on budget), Gantt chart, milestone.
5. Project closure: What was done well? What should be improved? What else did we learn? Archive documents.

---

* 5th edition (Newtown Square, PA: 2013).

SCIENTIFIC METHOD

In contrast to the intuitive thinking we discussed above, VA/VE is based on the **scientific method.** VA/VE follows a value-thinking process known as the value methodology. Originator Larry Miles referred to it as the "value job plan" and I call my version PISERIA. When Larry Miles, an engineer by profession, working in the purchasing department at General Electric, was assigned the task of developing the value methodology, it was only natural for him to come up with a language of reason and logical thinking that he learned from the scientific method, which involves the following five phases:

- Observation
- Problem definition
- Forming the hypothesis
- Testing the hypothesis (experimentation)
- Verification

The terms "value thinking" and "critical thinking" are used synonymously as both are based on the scientific method described above, following an intellectually disciplined method of diagnosis, analysis, and synthesis and including the commitment to apply reason and logic in the pursuit of excellence. The value methodology is presented in layperson's terms, following the steps of the scientific method, in Figure 6.1. Please refer to Appendices E through K for the agendas of each phase of PISERIA.

WHOLE SYSTEM THINKING

In the introduction of this book, I mentioned that the re-engineered VA/VE integrates the various value-improvement methodologies based on **whole-system thinking** in order to achieve enhanced results. Why is whole-system thinking important for the re-engineered VA/VE? (a) Because a systems thinking is the process of understanding how things influence one another within a whole. It enables you to see the big picture and appreciate interdependencies. (b) Because the whole-system approach is an effective method for solving problems from a holistic perspective, rather than as bundles of small isolated problems. (c) Because most of the

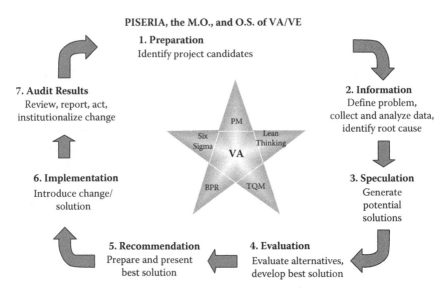

FIGURE 6.1
PISERIA, operating system of VA/VE.

problems we face in organizations, such as defects, mistakes, delays, and waste, represent a set of interrelated components in broader and more complex systems. (d) Because the focus on a silo approach of utilizing Six Sigma, Lean Thinking, Kaizen, and so on sub-optimizes the benefits of these value-improvement methodologies and undermines their impact on the rest of the system. (e) Because one lesson that nature teaches us is that everything in the world is connected to other things.

This concept is further corroborated by W. Edwards Deming's theory of profound knowledge, a management philosophy grounded in systems theory. It is based on the principle that each organization is composed of a system of interrelated processes and people that make up the system's components. Each organization, division, or department is an integrated and interrelated part of a larger system. Every part of the system affects the performance of other organizations in a larger system.

PISERIA, THE VALUE METHODOLOGY

Having provided this overview to Section II through reference to the related subjects of intuitive thinking, systems 1 & 2, inductive and deductive reasoning, DMAIC, the project-management process, and whole-system-thinking, we now come to an in-depth study of the integrative scientific method known as the value methodology, presented in the following Chapters 7 through 13. The circle diagram shown in Figure 6.2 symbolizes inclusion, unification, and wholeness as the various improvement methodologies are integrated in the seven steps of PISERIA.

If we have never done something before, we tend to be flying by the seat of our pants. But the PISERIA value methodology gives us a course of action, a flight plan. It is a value audit checklist. PISERIA is the battle plan for declaring war on waste (WOW) and improving performance. The battle plan includes the seven steps of the value-improvement process, capitalizing on the magical number 7 representing *completeness*. Perhaps the most important characteristic of VA/VE is this structured, disciplined, and organized approach of the PISERIA value methodology. It is a sequential, building-block process. If outstanding results are to be achieved in getting good, better, or best value, such a systematic plan of action is essential.

The diagram represents the thought processes of creative problem solvers that can be consciously learned by practicing on problems and

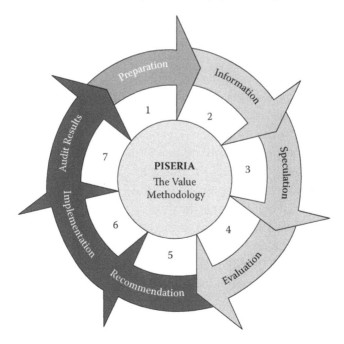

FIGURE 6.2
PISERIA: the value methodology.

opportunities, until the method becomes subconscious. As depicted in the green, yellow, and red colors of the PISERIA diagram, the value methodology is composed of three stages: ***the prologue*** (Preparation), ***the core of VA/VE*** (Information, Speculation, Evaluation), and the remaining phases of ***the aftermath*** (Recommendation, Implementation, Audit Results). The core elements of VA/VE form the Analyze-Create-Evaluate (ACE) step. They were referred to as the "Blast-Create-Refine" technique by Lawrence D. Miles. We blast away the structure to determine the function; we create alternative ways of accomplishing the function; and we refine one of the alternatives into a value-adding and cost-effective solution.

THE FIVE BASIC QUESTIONS

Before we delve into the "long form" of the seven-step value methodology, let us first review the "short form" known as the **Five Basic Questions.***

- What is it?
- What does it do?
- What does it cost?
- What else will do?
- What will that cost?

These five basic questions are the skeleton of VA/VE. A great deal of flesh and blood is necessary to bring them to life. They are the guideposts of VA/VE. The first three questions can be considered as problem-stating and the last two as problem-solving steps. The five basic questions will be explained in depth in Chapters 8 through 10.

Section II will include chapters devoted to each of the seven phases of the value methodology. A simple illustrative case study from a sugar corporation will be used to take you through the seven phases. In addition, the agenda of each phase is provided in Appendices E through K. Please refer to Appendix A, The Value Process Dashboard, for a summary of the seven steps of the value methodology.

---

* Lawrence D. Miles, *Techniques of Value Analysis and Engineering* (New York: McGraw-Hill, 1961), p. 18.

7

The Preparation Phase

As discussed in the overview, the first step of the scientific method is OBSERVATION. In VA/VE, that step is called the Preparation Phase. To observe, we use our five senses (some people have six), surveys, interviews, and questionnaires, which also involve the recording of data. Scientific instruments were developed to magnify human powers of observation, such as weighing scales, clocks, telescopes, microscopes, thermometers, and so on. For things that are unobservable by human senses, scientists use instruments such as indicator dyes, voltmeters, spectrometers, infrared cameras, and oscilloscopes.

In a court of law, a preliminary hearing is not a trial, but an effort to determine if crime has been committed. In VA/VE, the preparation phase is concerned with finding problems and challenges or recognizing opportunities. Opportunities are strengths we need to exploit while problems are weaknesses or threats that must be converted into opportunities for improvement. A soccer metaphor can also help to explain the differences between problems and opportunities: the goalie is in a *prevention focus* to prevent the problem or challenge of losing to a striker while the striker is in a *motivation focus* trying to find an opportunity to score a goal. Organizations as well as individuals should strive to build on their strengths and make their weaknesses irrelevant.

When we discussed the goals of VA/VE, we mentioned that the value methodology is a double-edged sword. There is a time when the opportunity edge is most effective. But there are other times when the problem-solving edge must do its part. Properly wielded, the two-edged sword of VA/VE can cut through the thickest of costs. The opportunity edge cuts before costs are incurred and the problem-solving edge cuts after costs have grown. According to John Naisbitt, "You don't get results by solving problems but by exploring opportunities because solving problems only restores the equilibrium of yesterday."

Let us review examples from a couple of industries about opportunity finding. In the leather industry, the hides of cows, goats, and sheep are converted to leather for shoes and garments. In the process, flesh remnants that result from the fleshing process become input for making fertilizers and glue. The shaving process to adjust leather thickness results in dust that is compressed into leather board to make suitcases. In the sugar industry, bagasse, the dry and dusty pulp that remains after the extraction of the juice from the crushed stalks of sugar cane, is used as a biofuel and in the manufacture of pulp and paper products and building materials.

Problem finding is more important than problem solving. While the analytical skills needed for problem solving are important, more crucial to managerial success are the perceptual skills needed to identify problems long before evidence of them can be found by even the most advanced management information systems. As John W. Gardner said, "Most ailing organizations have developed a functional blindness to their defects. They are not suffering because they cannot resolve their problems, but because they cannot see their problems."*

The Preparation Phase is like a mini-VA/VE or a prefeasibility study phase. Here are the tasks that we will perform during this first phase:

- Identify project candidates
- Perform SWOT analysis
- Perform Pareto Analysis
- Select priority project/triaging
- Define objective and scope
- Define project deliverables
- Determine project worth
- Select project worker(s)
- Complete project charter
- Approve project start
- Schedule the value effort

A clear identification of the problem to solve or the opportunity to exploit will be an output or work product generated by the aforementioned activities conducted during the Preparation Phase. In this project-planning phase, it's important to determine results to be produced and resources to be used, keeping in mind the mathematical formula of $Y = (f) X$.

---

* *Excellence* (New York: Harper & Row, 1961).

CANDY WRAPPER CASE STUDY

The case study that we will start with here and follow throughout the seven steps of the value methodology is the case of the Candy Wrapper. Figure 7.1 shows the product, the first strip having the surface inked 90%, the second inked 50%, and the third inked 25%.

Select VA Project and Project Worker(s)

At HVA (now the Ethiopian Sugar Corporation) 370 employees attended multiple Value Analysis seminars conducted over a 4-year period. The seminars were presented by this author. The employees who participated in the program ranged from senior managers, engineers, and departmental managers down to the foreman level. The seminars were conducted at the HVA estates of Wonji and Metahara during the factory stop and overhaul periods.

At the end of one of these seminars one of the participants, Bezabih Fetenaw,* selected "Candy Wrappers," a packaging example, as a project candidate for his value study.

Fetenaw's reasons for selecting the Candy Wrapper as his priority VA project were (a) the mounting cost of petrochemical products such as the

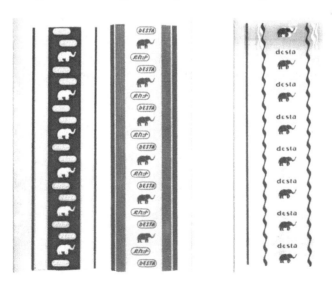

FIGURE 7.1
Candy wrapper.

---

* "Bezabih Fetenaw" is a pseudonym.

PVC plastic of which the wrapper is made, (b) the high volume of the material required annually (100–175 tons), and (c) the simplicity of the product for a first formal VA study (a good example of a low-hanging fruit to provide early wins for immediate impact). Details of this case study will continue on the following phases and chapters.

VA/VE Goal Setting

Unlike the traditional approach of setting 2% or 5% cost reduction goals, it is not logical or desirable to set VA/VE goals in terms of anticipated savings because (1) VA/VE is concerned with uncovering through research that which is not known and cannot be anticipated quantitatively in advance, and (2) when cost-reduction goals are arbitrarily established, they could be exceeded or failure to attain an arbitrary goal might be harmful to a reasonable VA/VE program.

Therefore, the goal setting that is recommended in VA/VE is a commitment to initiate VA projects.

Figure 7.2 shows a simple methodology (anonymous) for evaluating goal effectiveness.

Project Charter

According to the Project Management Institute's *Project Management Body of Knowledge* (PMBOK), "A project charter is a document issued by the

The SMART Way

- **S**pecific–It is the WHAT, WHY, and HOW. WHAT are your going to do? (A specific goal to lose 20 1bs in waistline). WHY is it important at this time? How are you going to do it?

- **M**easurable–If you can't measure it, you can't manage it. Establish a baseline, and set targets for improvement? GOALS THEORY teaches us that for goals to be effective, they need to be concrete and measurable.

- **A**ttainable or Agreed upon—Is the goal achievable? Is it a stretch goal? Is your aim to lose 20 1bs in one week attainable?

- **R**ealistic or relevant—Is it doable? Dose it relate to our business objective? Inspect what you expect.

- **T**imely or time Limited—Have we set a date for completion, phase by phase? A goal is a dream with a deadline. The time must be measurable, attainable and realistic.

FIGURE 7.2
Goal effectiveness method.

project initiator or sponsor that formally authorizes the existence of a project, and provides the project manager with the authority to apply organizational resources to project activities."* Value-improvement projects can be ranked and authorized by Return on Investment analysis. A project charter becomes a compass to keep the team firmly pointed toward the goals established when you started the journey. A good project charter becomes a daily reference point for settling disputes, avoiding "scope creep," judging the potential utility of new ideas as they arise, measuring progress, and keeping the development team focused on the end result. Terms of reference describing the purpose and structure of the project are usually part of the project charter.

SIPOC (Suppliers, Input, Process, Output, and Customers, discussed in Chapter 4) can assist in completing the project charter in numerous ways. The suppliers and customers are potential team members and/or stakeholders; the outputs are the metrics that will be used to measure the project; the inputs allow the project team to consider various potential critical areas; and, of course, the process itself provides the stop-start barriers.

A project charter will contain some or all of the following elements:

- Project name
- Objectives
- Goals
- Scope
- Deliverables
- Potential resources
- Cost estimates
- Critical success factors
- Assumptions and constraints (what is in or out of scope)
- Roles and responsibilities
- Key milestones
- Risks
- Business impact

Spend Analysis

Spend Analysis is the process of aggregating, cleansing, classifying, and analyzing corporate spending data for the purpose of reducing costs and

---

* Project Management Institute, *A Guide to the Project Management Body of Knowledge*, 3rd edn., (Newtown Square, PA: 2004), p. 368.

improving operational performance. It is a useful project identification and selection method. It can help you decide whether you want to prioritize on "low-hanging fruits" to provide small or early wins for immediate impact or use Pareto Analysis (the 80/20 rule) to focus on high-value targets.

Today, the CFO and senior management are valuing procurement or SCM unidimensionally, based on the amount of money the function can save because they understand that procurement is about saving, not spending. The spend analysis data in Figure 7.3 will help to answer questions like: What are you buying? Who are you buying from? Whose money are you spending? The visibility from that effort will provide you with actionable intelligence that will enable you to make strategic decisions.

Here is a list of some of the spend-analysis software providers that can help you slice and dice your spend data and enhance your value improvement opportunities: Ariba (SAP), Avotus, Basware, Bravo Solution, CVM Solutions, Emptoris (IBM), ePlus, Etesius, Fieldglass, Global eProcure (GEP), Iasta, Insight Sourcing Group (ISG), Ketera, Oracle, Procurian, ProcurePort, SAP, SAS Institute, SpendHQ, and Verian.

Value-Improvement Project Candidates

What have you done to weather the economic storm? When the market is asking for draconian cost reductions and your raw material costs (manufacturers' core spend) are spiraling out of control, where do you turn for savings? For most manufacturing companies, labor costs are only 5%–20%

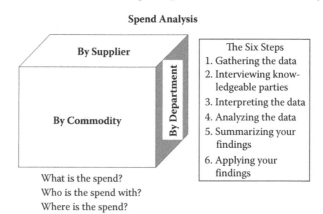

FIGURE 7.3
Spend analysis.

of total costs and overheads are fixed in the short term. How, then, can material and process costs be dramatically reduced without price concessions from suppliers? If your angle of attack is to enhance product value and process value, then the answer is VA/VE.

Here are the reasons why poor value or poor quality exists or continues to exist:

- Habits and attitudes
- Lack of technical information, cost information, time, ideas, or experience
- Personal prejudice
- Preconceived notions
- Temporary circumstances
- Political pressure
- Time pressure
- Job security (fear of personal loss)
- Honest wrong belief
- Risk of personal loss
- Desire to conform
- The N.I.H. factor (Not Invented Here)
- Authorship pride
- Resistance to change

Selecting projects with just a few obvious inputs or simply selecting the squeakiest wheel are not always the best methods. Figure 7.4 provides

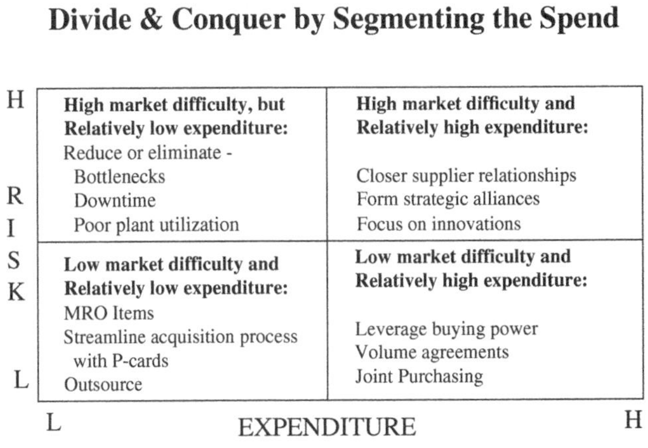

Divide & Conquer by Segmenting the Spend

H	**High market difficulty, but Relatively low expenditure:** Reduce or eliminate - Bottlenecks Downtime Poor plant utilization	**High market difficulty and Relatively high expenditure:** Closer supplier relationships Form strategic alliances Focus on innovations
	Low market difficulty and Relatively low expenditure: MRO Items Streamline acquisition process with P-cards Outsource	**Low market difficulty and Relatively high expenditure:** Leverage buying power Volume agreements Joint Purchasing

R I S K (label "RISK" runs vertically on the left axis; "L" at bottom of axis)

L   EXPENDITURE   H

FIGURE 7.4
Segmenting the spend.

some guidelines for identifying and selecting value-improvement project candidates as they cover the problems and opportunities from different perspectives.

If you are a value hunter, the examples in Figure 7.5 will help you identify some of the non-value–added operations in your organization. The potential VA/VE project will vary from company to company due to differences in the nature and complexity of each operation.

It is said that when Willie Sutton was asked why he robbed banks, he replied, "Because that's where the money is." Figure 7.6 shows four areas in any organization with a major procurement expenditure where VA/VE initiatives could have a significant financial impact:

1. What is your **corporate policy** regarding maverick or renegade purchasing, minority suppliers, and green purchasing?
2. How streamlined are your **governance structure** and **procurement process**?
3. How efficient and effective are your **procurement methods**?
4. How well leveraged are your organization's goods and services **commodity/category** spend?

If you are going to deploy valuable and scarce resources such as time and money, you should expect a payback for the effort. The PICK Chart in Figure 7.7 can help select the project that will provide the biggest benefit in

Non-Value Added Operations: Supply Chain Challenges

- Machine breakdowns
- Tools unavailable or broken
- Late delivery of materials
- Poor physical layout
- Poor workflow
- Poor quality control
- Wrong/defective raw materials
- Parts/materials shortages
- Unaware of work assignments
- Poor scheduling methods
- High material waste
- Low equipment utilization

- Supervisors not supervising
- Weak employee training
- Unknown performance requirements
- Poor work habits
- Late starts/early quits
- Employee self-pacing
- Leaving work stations
- Idle due to lack of materials
- Unbalanced work load
- Double handling
- Emphasis on correction, not prevention

FIGURE 7.5

Non-value-added operations. (From Bhote, K., *The Power of Ultimate Six Sigma*, AMACOM, New York, 2003, p. 217).

Procurement Cost Impact

Corp.
Policy

Procurement Process

Procurement Methods

Commodity/Category Management

FIGURE 7.6
Procurement cost impact.

Selecting a Project (The PICK Chart)

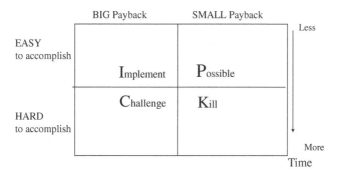

FIGURE 7.7
P.I.C.K. chart to select project. (From Carreira, B. and Trudell, B., *Lean Six Sigma that Works*, AMACOM, New York, 2006, p. 65; George, M., *Lean Six Sigma for Service*, McGraw-Hill, New York, 2003, p. 293.)

the shortest amount of time. You would place the projects on the matrix according to how easy or hard they will be to implement, how long it would take to complete them, and how big the return on investment would be. Like triaging in a hospital, that is, sorting out patients to determine priorities for treatment, you would prioritize based on factors such as the project being easy to implement, fast to implement, inexpensive to implement, or within the team's control.

Red Flags for Project Selection

With finite resources and infinite project possibilities, project selection could be the most important step in the project life cycle. Early termination of a troubled project may be worth consideration. In *The Art of War* ancient Chinese general Sun Tzu wrote that the best strategy for winning a war is avoiding it in the first place. The next-best time to quit is as soon as possible and preferably before sunk costs make the decision to quit a troubled project increasingly complex and difficult. Here is a list of red flags for project selection*:

- Lack of strategic fit with mission
- Lack of stakeholder support
- Unclear responsibility for project risks
- Risks outweigh potential benefits
- Unclear time component
- Unrealistic time frame, budget, and scope
- Unclear project requirements
- Unattainable project requirements or constraints
- Unclear responsibility for project outcomes

Tollgate Review is a formal review at the end of each phase of the value methodology (PISERIA) to approve passing on to the next phase, to monitor progress and keep projects on track. It is at this stage that a view is formed as to whether subsequent phases are likely to yield sufficient value to justify the cost of the study within the terms set. That means that monitoring and evaluation need to take place prior to moving to the next phase.

---

* Joni Seeber, Project Selection Criteria: How to Play It Right. American Society for the Advancement of Project Management, 2011. http://www.org/asapmag/articles/SelectionCriteria.pdf.

8

The Information Phase

The major aspects of problem solving involve **Recognition**, **Definition**, and **Solution**. Recognition was covered under the Preparation Phase; definition is the subject matter of this Information Phase; and solution will be the next chapter under the Speculation Phase.

CANDY WRAPPER CASE STUDY

We will continue our case study of the Candy Wrapper, begun in the last chapter in the Preparation Phase. The objectives of the Information Phase are to collect and analyze data in order to competently define the problem of the candy wrapper, as shown in Table 8.1.

Note that the unit cost climbed from $2200 per ton to $3850, meaning a 75% increase over the 4-year period due to the rising cost of oil.

Even though the plastic material represented a higher percentage in expenditure than the inking, the project worker preferred for his first study to focus on the inking portion, as it was more illustrative of LACK OF VALUE SENSE in its design. Moreover, this is a small foreign purchase item draining a lot of foreign exchange. The project worker has now defined the problem.

WHAT IS IT?

The Information Phase is a discovery process and an understanding phase. A doctor wouldn't prescribe medicine before knowing what is ailing the patient! To be familiar with a product we need to identify the item by name, part number, manufacturer, size, shape, usage, quantity, lead-time, material specification, and method of attachment, and if

TABLE 8.1

Candy Wrapper Case Study: Information Phase

What Is It?	What Does It Do?
PVC Clear Film	Contains Candy
	Protects Candy
INKING (90%)	Provides beauty
	Attracts customers
	Prevents consumer from seeing candy
	Identifies product (DESTA)
	Identifies maker (via logo)

	Cost per Ton in U.S.$				
What Does It Cost?	Year 1	2	3	4	%
PVC Clear Film	1320	2040	2214	2310	60
90% Inking	880	1360	1476	1540	40
Total	$2200	3400	3690	3850	100

possible, obtain drawings and a sample. Since a problem well defined is half solved, we will apply knowledge and skills and use different tools, such as value-stream mapping and root-cause analysis, separate symptoms from problems and facts from opinions, distinguish between reasons and excuses, and use other tools like Fishbone and the five Whys. The key to a good problem definition is ensuring that you deal with the real problem, not its symptoms. You may mistakenly assume that the poor performance of your department is a problem with the individuals submitting work, while the real problem could be lack of training or inequitable distribution of work. Therefore, attack the problem and not the person.

WHAT DOES IT DO?

Answering the first question of WHAT IS IT? establishes the parameter of the study. The next question is WHAT DOES IT DO? The answer to this question provides functional definition. The functional approach is the heart of VA/VE and lends this discipline the unique character that distinguishes it from other problem-solving methods. It relates cost to function while other methods relate cost to product or part. Therefore, for VA/VE, the basic reality is the function; the part or the product is not the design, and the design is not the function. Function is what makes the product work or sell (i.e., solve the problem). VA/VE is interested in the functional sense of what the product does, not in the structural sense of what it is. Remember what Theodore Levitt told us: "People don't buy ¼" drill bits; they buy ¼" holes."

The Anatomy of Function

Function Definition: Two-Word Description of Function as **VERB and NOUN**

If VA/VE is to give the best results, function must be boiled down to its simplest two-word expression as an *active verb* and a *measurable noun*. To state what something does in two words is not easy, but it helps to simplify terminology and create better understanding. For example, a carpenter's hammer drives nails; a saw cuts wood; a hacksaw cuts metals.

Two major advantages of the verb-noun description are (1) it pinpoints the function to be performed, keeping in mind that a problem clearly stated is already half solved, and (2) it provides for effective brainstorming for alternative solutions—because when functions are defined in this way, one can almost automatically suggest other ways of performing the function—and it enables designers to visualize their goal without being restricted by a previously conceived design.

Kinds of Functions (Function Identification)

Customers purchase a product or service because it will provide certain functions at a cost that they are willing and able to pay. Expenditures that increase the functional capability of an item beyond that which is needed are of little value to the user. Thus, anything less than the necessary functional capability is unacceptable and anything more is unnecessary and wasteful.

A product must have a basic function and one or more secondary functions. Together they compose the essential function. The basic function performs the primary duty of a product or service; it is the reason for its existence.

Secondary functions are features that result from a specific design approach taken to satisfy the basic function. Secondary functions may be categorized into three major classifications: required, aesthetic, and unwanted. For an example, Figure 8.1 shows the application of FAST diagramming to a pencil.

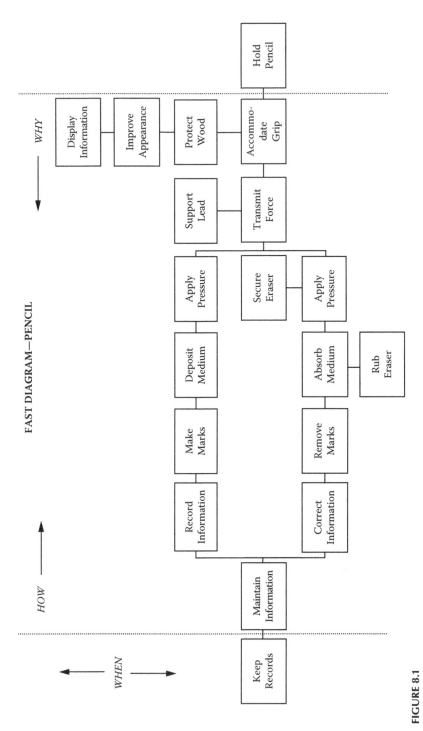

FIGURE 8.1

The FAST diagram. (Adapted from an example developed by J. Jerry Kaufman.)

Purpose–Function–Method

In doing function definition and problem analysis, it is necessary to distinguish among these three concepts:

1. PURPOSE: The objective, the reason *why* something is being done or is to be done.
2. FUNCTION: The action, *what* is being done or is to be done in order to achieve the purpose.
3. METHOD: The system, *how* the function is being performed or is to be performed.

For example:

PURPOSE: Prepare food
FUNCTION: Heat material (bread)
METHOD: Electric toaster

FUNCTIONAL ANALYSIS SYSTEMS TECHNIQUE

Classical FAST Model: Functional Analysis Systems Technique (FAST) builds on the VA verb–noun pair analysis. It is a graphical mapping tool displaying the interrelationship of functions to each other in a "how–why" logic. This was developed by Charles Bytheway at Univac in 1964.

Hierarchy Function Model: A vertical hierarchical chart of functions. This places the basic function at the top. The function of each major system is placed beneath the basic function. The functions that support each of these functions are then placed on the next row. This process is continued until the team feels the level of detail is sufficient for the intent of the study.

Technical FAST Model: A variation to the Classical FAST Model that adds "all the time" functions, "one time" functions, and "same time" or "caused by" functions.

Customer-Oriented FAST Model: This variation of the FAST diagram was developed to better reflect that it is the customer who determines value in the function analysis process. Customer-oriented FAST adds supporting functions: to attract users, satisfy users, assure dependability, and assure convenience. The project functions that support these customer functions are determined by using the how–why logic.*

---

* SAVE International Study Guide.

The twin questions of What does it cost? and What is it worth? can be explained better with the classic example of the **pencil** that is comprised of materials (wood, lead, rubber, metal, and paint) purchased from different sources. The basic part of the pencil is the lead/graphite and all others are secondary parts. When you observe the cost/worth ratio of 0.07–0.02 in Figure 8.2, you can appreciate the potential for value improvement. The worth of each function is the estimated cost of the least expensive way to fulfill that function of "makes marks," which is two cents. The comparison indicates whether the study should be terminated because cost and worth are approximately equal, or pursued because cost greatly exceeds worth. A fundamental tenet of a value methodology is that basic function (the necessary purpose of the project) must be preserved. This is because the basic function reveals the usefulness of the project and the reason for its existence.

If a whole regular pencil is to be used at a 20 or 40-h seminar where participants are supplied with a workbook that contains all of a speaker's presentation slides, there will be almost no need to erase notes taken by participants. If the eraser is eliminated, there will be no need for the ferrule. If participants received meeting materials, then they will need only half a pencil, like the ones you find at golf courses.

For in-depth information about FAST Diagramming, please refer to *Functional Analysis: The Stepping Stones to Good Value*, by Thomas J. Snodgrass and Muthiah Kasi (University of Wisconsin System, 1986);

We've got the
magic formula!

$$V = \frac{F}{C}$$

Function and Cost Analysis of a Pencil

ITEM	FUNCTION	COST
Pencil	Makes marks	.07
Eraser	Removes marks	.01
Ferrule	Secures eraser Anchors eraser	.005
Wood	Holds lead Displays info	.025
Paint	Protects wood Provides color	.005
Markings	Identifies product	.005
Lead	Makes marks	.02

FIGURE 8.2
Pencil.

Stimulating Innovation in Products and Services: Function Analysis and Function Mapping, by J. Jerry Kaufman and Roy Woodhead (Wiley-Interscience, 2006); *Function Approach to Transportation Projects: A Guide to Value Engineering,* by Muthiah Kasi (iUniverse), 2009.

VALUE-STREAM MAPPING

Value-Stream Mapping (VSM) is a lean thinking methodology used to evaluate company operations, eliminate waste in its many forms, and substantially streamline business processes from the customer to the supplier. VSM is a structured process that captures the flow of product, people, tools, resources, and instructions. VSM should be a collaborative process regardless of whether it is managed on a brown paper bag or with Post-it notes, or using automated process-modeling tools. The goal is to identify value in the eye of the customer and eliminate all non-value–added activities.

A value stream is all activities that create value. It starts with raw materials or initial information and ends with the customer or end user. Wherever there is a product or service for a customer, there is value stream.

A **process map** (Figure 8.3) is an organized visualization of all the interrelated activities that combine to form a process. Don't get hung up

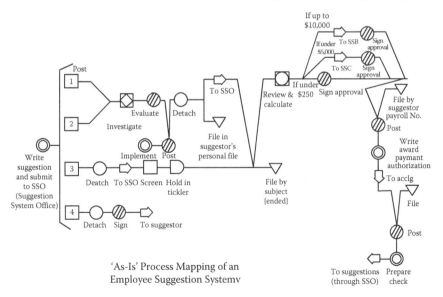

'As-Is' Process Mapping of an
Employee Suggestion Systemv

FIGURE 8.3
Value stream map of a current state.

on symbology. At Toyota, where the technique originated, it is known as "material and information flow mapping." Use Post-it notes for your data and put them on an easel chart or white board.

ROOT-CAUSE ANALYSIS

In the Information Phase, we collect and analyze data as well as identify root causes and constraints. Root-Cause Analysis (RCA) is one of the problem-solving methods that is aimed at identifying the root causes of problems. RCA is often considered to be an iterative process, and is frequently viewed as a tool of continuous value improvement. Examples of RCA techniques are the Cause and Effect Diagram (Fishbone Diagram) shown in Figure 8.4, The Five Whys, and Pareto Analysis.

THE FIVE WHYS*

The five Whys is a simple problem-solving technique to get at the root of a problem quickly. It was developed in the 1970s by Taiichi Ohno, Toyota's chief production engineer. The five Whys involves looking at a problem and

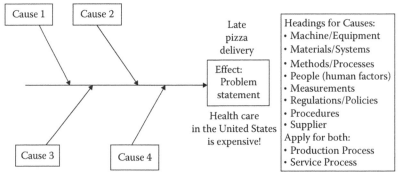

The Cause and Effect Diagram

a.k.a. Ishikawa or Fishbone Diagram

Brainstorm using post-it notes to collect data; Analyze data to determine the root cause.

FIGURE 8.4
The cause and effect diagram.

---

* James Womack, Daniel Jones, and Daniel Roos, *The Machine that Changed The World: The Story of Lean Production* (New York: Harper Perrenial, 1990), p. 57.

asking, "Why?" and "What caused this problem?" Quite often, the answer to the first "why" will prompt another "why" and the answer to the second "why" will prompt another and so on; hence the name of the strategy. Here's an example:

My car won't start. Why?
The battery is dead. Why?
The alternator isn't charging. Why?
The fan belt is cut. Why?
A rock hit the belt. Why?

WHERE DO WE GET THE INFORMATION?

We gather information from the best sources. To save yourself and others time and trouble, it is imperative that you contact people who can give you the information you need accurately and quickly. Use industry specialists to extend specialized knowledge. Solicit the aid of others.

HOW DO WE GATHER AND RECORD INFORMATION?

Search methods are "Look, Listen, Learn" tours. A suitable search method is like a large magnet picking up a needle lost in a haystack. **Fact-Gathering Techniques** include:

Interviews
The Internet
Surveys
Plant visits
Personal observation
The Columbo approach
Checklisting
   What is done?   Why is it done at all?
   Where is it done?   Is it done there?
   When is it done?   Is it done then?
   Who does it?   Does this person do it?
   How is it done?   Is it done this way?

The practicality and quality of your value study will depend directly on how thorough you are during the information phase. Following are some more guidelines for the information phase checklist.

Specifications

Are the specifications realistic? Are the specifications required by the customer or are they guidelines only? What are the desired life and reliability requirements?

Function

What does the product or service do? Is the function a needed secondary function or an imposed secondary function? What does it do unnecessarily? Can this function be eliminated? Have functions been separated into work and sell?

Design

What alternative designs were considered? Why were alternatives rejected? In what stage of maturity is the design?

Special Requirements

Is a severe environment involved? Are there special requirements relative to installation? Maintenance? Testing? Safety? Are special treatments, finishes, or tolerance required?

Materials

Are special, hard-to-get, or costly materials specified? What alternative materials were considered? Why were they rejected? Are the materials used difficult to handle, process, or work? Are they hazardous?

Manufacturing

How are the component parts made? Why are they made that way? Who makes them—supplier or in-plant? What will be the economic order quantity (EOQ)? What are potential sources? What methods, machines, and processes are used?

Rudyard Kipling's poem "*I Keep Six Honest Serving Men*" reminds us of how to gather information: "I keep six honest serving-men. (They taught me all I know.) Their names are What, and Why and When and How and Where and Who."

Fact Recording Techniques

Fact recording techniques include note taking, flow-charting, process mapping, and FAST diagramming.

WHAT DOES IT COST?

Price is what you pay for the product or service. To understand what makes up a price, refer to Figure 8.5. **Cost** is what you spend to make a product or

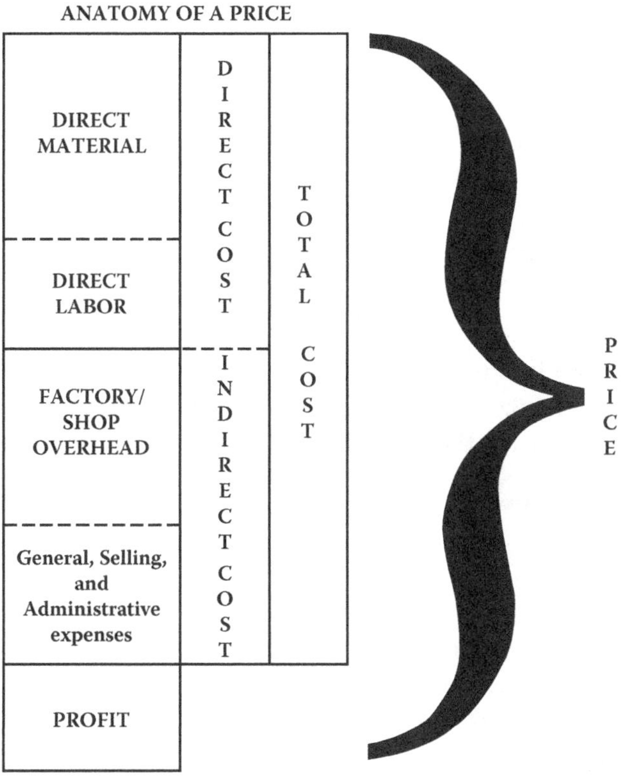

ANATOMY OF A PRICE

FIGURE 8.5
Anatomy of a price.

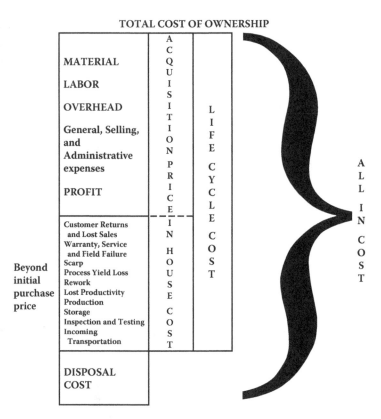

FIGURE 8.6
Total cost of ownership.

provide a service. Cost is a design parameter because it controls resource expenditure. To find out what the Total Cost of Ownership is made up of, refer to Figure 8.6. **Value** is what you get, that is, the satisfaction the customer derives from the function performed by the product or service.

DOES COST DETERMINE PRICE, OR VICE VERSA?

Peter Drucker wrote, "The shift from cost-led pricing to price-led costing in which the price the customer is willing to pay determines allowable costs, will force companies into economic-chain costing."* If the only sound way to price is to start out with what **the market** is willing to pay, then we ought to design products to that price specification.

---

* The information executives truly need, *Harvard Business Review*, January–February. 1995.

Therefore, if selling price is truly set by the marketplace, and if Profit equals Selling Price minus Cost, then we ought to be able to find ways and means to reduce cost or eliminate unnecessary cost (waste).

In essence, VA/VE is not about cost cutting; it is an über strategy to spend money wisely and purposefully. Remember that cost is a design parameter because it controls resource expenditure to perform functions required by the customer. Given here is a sample of how we can look at cost from different perspectives.

Cost Arrangement Analysis: As indicated in Figure 8.5 the elements of the anatomy of a price are arranged as material, labor, overhead, G, S, & A, and profit.

Cost Allocation Analysis: Costs could be allocated per dimension (length, area), weight, time, and requirements.

Cost Visibility Analysis: Cost could be made visible with a pie chart, tree chart, or bar chart.

Life Cycle Costing: As indicated in Figure 8.6, LCC goes beyond the initial purchase price and includes the in-house cost.

Design To Cost: Establishes, as a design goal, a unit production cost that you can afford to pay (for the quantities you need) as a primary design parameter (equal with performance) in order to assure production at a limited prespecified unit production cost.

Zero-Based Budgeting: Customarily, those in charge of an established budgetary program are required to justify only the increase sought above last year's appropriation. Under a zero-based budgeting plan what a manager is already spending is not accepted as a starting point and the manager will be required to present different ways of performing the same activity with realistic and meaningful costs. In the vernacular of VA/VE, what else will do and what will that cost?

Break-Even Analysis: Determines the point at which revenue received equals the costs incurred to generate the revenue. Here is an example how to calculate the break-even point (BEP) for a 3-day seminar.

> **Variable Cost** (VC): $16 for handout materials and $49 for 3 days of lunch and break refreshments per attendee. Total VC equals $65.
> **Seminar fee (price)**: $650.
> **Contribution Margin** (CM): Seminar fee of $650 minus VC of $65 equals CM of $585.
> **Fixed Cost** (FC): Printing, instructor's fee $4550.

Break-Even Point: Divide fixed cost by contribution margin. FC of $4550 divided by CM of $585 equals BEP of eight attendees. It will take fees paid by eight attendees to cover all expenses.

For another example, it is estimated that Boeing needs to sell a whopping 1,500 Boeing 787 Dreamliners before reaching its breakeven point, which may be around the year 2021.*

---

* When is Boeing's 787 Dreamliner Breakeven Year? 24/7 Wall Street, September 26, 2011. http://247wallst.com/2011/09/26/when-is-boeings-787-dreamliner-breakeven-year-ba/.

9

The Speculation Phase

In the Speculation Phase, we move from "matters of fact" (Information Phase) to "matters of value," from "what it is" to "what it should be." In other words, the question we are trying to answer here is, "What else will do the job?" That is, what other ideas can we think of that will perform the same function as the item under study, keeping in mind that our focus is on the function and not what the item is.

To continue with the Candy Wrapper case study, after collecting and analyzing data in the Information Phase and noting the rising cost of the wrapper over 4 years, the project worker thought about other ways to design the wrapper.

CANDY WRAPPER CASE STUDY

"What else will do?"

- What other designs can perform the present functional requirements more effectively?
- What other designs can provide additional functional benefits?
- What other lower cost design can be created?

Upon request by the project worker, the current supplier submitted the following six alternative designs:

	% INKING
Design #1	75
Design #2	50
Design #3	25
Design #1A	25
Design # 2A	25
Design #3A	25

THE ALTERNATIVE COURSES OF ACTION: WHAT ELSE WILL DO?

There are five types of action to choose from when evaluating alternative courses to arrive at an appropriate solution. Each type has a different purpose:

1. **Interim action** buys the manager time for finding the cause of a problem.
2. **Adaptive action** lets the manager live with the tolerable effects of a problem or with an ineradicable cause.
3. **Corrective action** gets rid of the known cause of a problem.
4. **Preventive action** removes the possible cause of a problem or reduces its probability.
5. **Contingency action** provides stand-by arrangements to offset or minimize the effects of a serious potential problem.

It is not what happens to you, but how you react to it that matters.

—**Greek philosopher Epictetus**

In the Speculation Phase, in its search for solutions to problems, the value methodology puts to use, as applicable, the following performance-enhancing tools—because supply-chain management also has its **SCM toolkit** just as doctors and mechanics have toolboxes.

Standardization	Benchmarking
Hedging	Total cost of ownership
Contract negotiation	Cycle-time reduction
Competitive bidding	Material substitutions
Supplier rationalization	Volume leveraging
Systems contracting	Cross-functional sourcing
Supplier-managed inventory	Product design improvements
Consignment buying	Investment-recovery initiatives
Global sourcing	Group purchasing
E-procurement	Procurement card
Outsourcing	Countertrade

CREATIVITY IN THE SPECULATION PHASE

The Speculation Phase is the creative-thinking phase. It is said that the natives of Fiji carry water in empty Standard Oil cans, and clothe their infants in Pillsbury flour bags (Samuelson, p. 52).* This, to an economist, means something else, but to the value analyst, it is creativity par excéllence! In his book about how to produce ideas, James Webb Young writes, "With regard to the general principles which underlie the production of ideas, it seems to me that there are two which are important, (a) An idea is nothing more or less than a new combination of old elements, and (b) The capacity to bring old elements into new combinations depends largely on the ability to see relationships."†

A rich tradition of creativity education was born at SUNY Buffalo State. Sowing the seeds for the study of creativity, in 1948 Alex F. Osborn published *Your Creative Power: How to Use Imagination*,‡ in which he introduced the concept of brainstorming. Then in 1953 he published *Applied Imagination: Principles and Procedures of Creative Problem-Solving*.§ Osborn's dream was fully realized when Dr. Sidney Parnes and Dr. Ruth Noller established a permanent academic home for the Creative Studies Program at Buffalo State.

---

* Paul A. Samuelson, *Economics: An Introductory Analysis*, 7th edn. (New York: McGraw-Hill, 1967), p. 52.
† *A Technique for Producing Ideas* (N.p.: Important Books, 2012).
‡ New York: Charles Scribner's Sons.
§ New York: Charles Scribner's Sons.

Then Dr. Noller developed her creativity formula, which we can express roughly as Creativity = Knowledge x Imagination x Evaluation. While we accept that as the total definition of creativity, however, in VA/VE we separate evaluation from the Speculation Phase (knowledge and imagination) in order to get an abundance of ideas without being hindered by judgment and criticism. Therefore, in VA/VE, we first IDEATE, then EVALUATE. That means our thinking process is mainly twofold: (a) a creative mind (knowledge and imagination) that visualizes, foresees, and generates ideas, and (b) a judicial mind (Evaluation) that analyzes, compares, and chooses.

Whether we examine the high level of conceptual thinking concerned with national defense strategy, or the technical problems of overhauling a jet engine, or such human problems as to how to overcome resistance to change, there are four common mental skills that are learned and that can be applied and further developed:

1. PERCEPTION: the ability to observe and apply attention. It is sensing the existence and characteristics of anything by seeing, hearing, tasting, smelling, or feeling.
2. RETENTION: the ability to memorize and recall. It is putting a bit of information into memory or taking it out, whether intentionally or not.
3. REASONING: the ability to analyze and to judge. It is relating things, to make decisions, judgment, or deductions and inductions; going from what is known to what is desired, according to some learned pattern with measurable, logical criteria.
4. CREATIVITY: the ability to visualize, foresee, and generate ideas. Putting the pieces together in new ways, or with something new added, to meet constantly changing needs.

Through the first two, WE LEARN. Through the latter two, WE THINK.

Who is creative?—Small children, pioneers, writers, and so on.

What motivates creativity?—Discontent, wars, fear, necessity, curiosity, and so on.

What inhibits creativity?—A malady named by value engineers as "mental blockitis" limits the use of our creative ability. Yet, creativity is the means by which we will have to obtain the answer to the question, What else will do? If we wish to succeed in removing poor value and create instead good value, we have to overcome the following four common mental blocks:

1. **Cultural Blocks**
   - Pressure to conform to "proper" patterns or customs at home, in school, and at work
   - Overemphasis on cooperation or competition
   - Too much faith in statistics
   - Overdependence on generalizations
   - All-or-nothing attitude
2. **Emotional Blocks**
   - Fear of making a mistake or making fools of ourselves
   - Playing it safe and not making optimum decisions
   - An overriding desire to succeed quickly
   - Lack of self-confidence, distrust of associates, and fear of authority
   - Inability to relax and let "incubation" take place
3. **Habitual Blocks**
   - Failure to develop the best habit of perception
   - Being a conformist
   - Self-protection or security
   - Using tried-and-true procedures only
4. **Perceptual Blocks**
   - Failure to use all the senses for observation
   - Inability to define terms
   - Failure to distinguish between cause and effect
   - Difficulty in isolating the problem

To achieve good value, the existence of these four mental blocks must be recognized and replaced by creative flexibility.

CREATIVE-THINKING TECHNIQUES

Several creative-thinking techniques are used in the Speculation Phase. They may be used individually or in combination, depending on the project under study. Given here are some of the more widely used techniques.

Brainstorming

We all have a "gold mine between our ears," and one of the best tools for extracting that gold is brainstorming. The advertising industry originated brainstorming as a technique to stimulate creative thinking on advertising copy. Its applications broadened after Alex Osborne of the big advertising firm of Baten, Barton, Durstine & Osborne popularized the technique with his books *Your Creative Power* and *Applied Imagination.*

The free association of ideas is the hub on which brainstorming turns. Here are the rules of the game, adapted from Osborne:*

- Judgment is ruled out: Do not evaluate.
- Quantity is wanted: Quantity breeds quality.
- Freewheeling is welcome: Remember, they said man couldn't fly.
- Seek combination and improvement: Cross-fertilize.
- Record all ideas rapidly: The flip chart works well.
- Set a deadline: Limit meeting time.

Affinity Diagram

Also named the KJ method after its developer Japanese anthropologist Kawakita Jiro, an affinity diagram helps to synthesize large amounts of data by finding relationships among ideas. Here's how it works:†

PURPOSES and BENEFITS

- Allows a team to creatively generate a large number of ideas and then organize them into groupings based on their natural relationships.
- Efficient tool for gathering and grouping of like ideas.
- Encourages creativity.
- Breaks down long-standing communication barriers.
- Encourages nontraditional connections among ideas.
- Allows breakthroughs to emerge naturally.
- Encourages ownership of results that emerge because the team creates both the detailed input and general results.
- Overcomes team paralysis brought on by an overwhelming array of options and lack of consensus.

---

* Osborn, *Applied Imagination,* pp. 300–301.
† Adapted from Michael Brassard, *The Memory Jogger: A Pocket Guide of Tools for Continuous Improvement* (Salem, NH: GOAL/QPC, 1985), pp. 12–13.

PROCEDURE

- Follow guidelines for brainstorming.
- Record each idea on Post-it note by using verb and noun.
- Sort ideas into related groupings; develop headers for the groups.

Synectics

Synectics, a Greek word, means the joining together of different items and reflects the early discovery of the synergistic effects of "banging things together." The creative equation of A + B = C illustrates how new ideas can be created from two old ideas.

Synectics as a creativity technique means the integration of diverse (heterogeneous) individuals into a problem-stating and problem-solving group that results in artistic or technical inventions.

Synectics uses analogies. In testing and observing 3000 executives over a 6-year period, professors Jeffrey Dyer, Hal Gregersen, and Clayton Christensen* noted five important discovery skills for innovators: associating, questioning, observing, experimenting, and networking. The most powerful overall driver of innovation was associating—making connections across "seemingly unrelated questions, problems, or ideas." For instance, the automobile made the horse and carriage obsolete and the four legs evolved into four wheels.

The synectics process involves making the strange familiar and the familiar strange. Making the strange familiar means understanding the problem to be solved, converting its strangeness into familiarity by using our favorite VA/VE questions, like What is it? What does it do? What does it cost? Making the familiar strange is to distort, invert, or transpose the everyday ways of looking at things by use of analogy, free association of ideas like *similarity*, *contiguity*, and *contrast*. Synectics research has identified the following four mechanisms (psychological tools), each metaphorical in character, for making the familiar strange†:

Personal Analogy
    If I were the customer what would I buy?

---

* The innovator's DNA. *Harvard Business Review*, December 2009.
† William Gordon, *Synectics: The Development of Creative Capacity* (New York: Harper & Row, 1961).

Direct Analogy
    The bird and the airplane
Symbolic Analogy
    Using images to represent elements of a problem
Fantasy Analogy
    Wish fulfillment

Checklisting

A checklist is a "think list" to spur more creative ideas for using an item or to serve as idea needlers or idea clues or leads. Various checklisting methods are listed here.

Tests for Value (GE): A now classical checklist developed by General Electric
- Does using the product contribute VALUE?
- Is its cost proportionate to its usefulness?
- Does it need all of its features?
- Is there anything better for the intended use?
- Can a usable part be made by a lower-cost method?
- Can a standard part be found that will be usable?
- Is it made on proper tooling, considering quantities used?
- Do material, reasonable labor, and overhead total its cost?
- Will another dependable supplier provide it for less?
- Is anyone buying it for less?

Checklists Galore!
- ELIMINATION: waste, duplication, spoilage
- SUBSTITUTION
- COMBINATION: forms, operations, functions
- STANDARDIZATION
- SIMPLIFICATION: work, products, methods
- REDUCE, RECYCLE, REUSE: sustainability
- CHANGE OF PLACE, SEQUENCE, PERSON
- IMPROVE: quality, human factors, performance
- INCREASE: safety, sales, productivity
- SAVE: time, money, material
- REDUCE: costs, scrap, weight

The 5S + 1 for the Visual Workplace*

A Japanese methodology for workplace organization developed by Hiroyuki Hirano. Both the Japanese and English words begin with *s*:

Sort (*Seiri*) = Remove all unnecessary materials and equipment.

Straighten (*Seiton*) = Arrange and place things so that they can be easily reached and with the shortest path.

Shine (*Seiso*) = Clean and keep machines and working environment clean, no debris or dirt in the workplace.

Standardize (*Seiketsu*) = The state of cleanliness or organization—Maintain and monitor the first 3 S's.

Sustain (*Shitsuke*) = Discipline, maintain the practice that has been established in the above four steps.

Safety (often called "6S" or "5S + 1") = Safety commitment.

Benchmarking

Benchmarking is a process of MEASURING your practices, COMPARING them with best-in-class, and continuously IMPROVING performance toward best-in-class status.

Benchmarking Steps:

1. Determine what to benchmark
2. Select benchmark team(s)
3. Evaluate internal performance
4. Select benchmark target organization(s)
5. Prepare site visit preparatory package
6. Complete preparation activities
7. Conduct interview-team training
8. Conduct benchmark visit
9. Debrief interview team as soon as possible after visit
10. Complete final deliverables

---

* Hiroyuki Hirano, *5S for Operators: Five Pillars of the Visual Workplace* (New York: Productivity Press, 1996); Carreira and Trudell, *Lean Six Sigma*, pp. 135–148; Michael George, *Lean Six Sigma for Service* (New York: McGraw-Hill, 2003), p. 302.

Best practice is past practice! Benchmarking is just a tool for catching up, not for jumping way ahead. As Michelle Bechtell expressed it, "When we do nothing more than benchmark others, we limit ourselves to implementing the best of today's known management practices. This can provide valuable industry information, but it generates reactive, not proactive management strategy."*

---

* *The Management Compass: Steering the Corporation Using Hoshin Planning* (New York: AMACOM, 1995).

10

The Evaluation Phase

CANDY WRAPPER CASE STUDY

In the Speculation Phase the project worker determined alternative designs for the candy wrapper to meet the functional requirements of the product with the goal of improving performance and reducing cost. In this Evaluation Phase, the project worker analyzed the merits of each design with respect to function and cost.

Function Evaluation

Removing the excessive decorative inking from the center of the wrapper enabled exposure of the product (candy) through the transparency of the film. Customers can now choose their preferred flavor guided by the color of the candy.

The company logo (elephant) and the product name (DESTA candy) having been changed from white ink to bold shades of red, green, and orange, the candy can now be seen more vividly.

Beautification of the wrapper was achieved by using strong colors of green, red, and orange instead of white at the edges of the wrapper. Furthermore, the color of the candy itself enhanced the beauty of the overall package.

The 20% and 10% inking options were dropped because that amount of ink was not sufficient to provide the required degree of beauty. Hence, the 50% and 25% inking were preferred.

Cost Evaluation

Six alternative designs were compared against the 90% inking.

	Stage 1			Stage 2	
Prototype Sample	Inking Content	Price Reduction	Prototype Sample	Inking Content	Price Reduction
Design #1	75%	6%	Design #1A	25%	24%
Design #2	50%	17%	Design #2A	25%	24%
Design #3	50%	17%	Design #3A	25%	24%

Obviously, the 50% and 25% inking offered the highest cost-reduction potential, with 17% and 24% lower costs, respectively.

WILL IT WORK? WHAT WILL IT COST?

Following the logical steps of the value methodology, the objective of the Evaluation Phase is to analyze the creative ideas generated in the Speculation Phase by screening the various alternatives for their feasibility (Will it work?) and cost effectiveness (What will it cost?), and then to select the best idea for further refinement. In other words, the Evaluation Phase involves technical considerations for suitability and economic considerations for cost and availability. Functions are evaluated by comparing the relative costs of alternate methods of performing the function. Cheapening (reduction of function) is prohibited as a solution and over-engineering is expensive.

The key questions to be addressed are: Will each option perform the basic function? Will it work? Will it meet requirements? What does each alternative cost? Which is the least expensive alternative? What is it worth, that is, what is the simplest, lowest-cost method that will perform the function? This then is what is meant by relating the elements of product worth to their corresponding elements of product cost.

Risk identification and analysis is integrated into the Evaluation Phase to examine the performance of alternatives. Risk management includes maximizing the probability and consequences of positive events and minimizing the probability and consequences of events adverse to the project objective. Here are some pointers for evaluating Technical and Economic Risks.

Questions about Technical Risk

- Will the idea work as planned?
- Can safety hazards be kept within prescribed limits?
- Will the product be as dependable as intended—available when needed, reliable, and repairable?
- Will the product last as long as expected?

Questions about Economic Risk

- Loss of opportunity: Will the effort and resources spent on the proposed product or change deprive a more worthwhile option?
- Cash flow: Will resources consumed in launching this product or making this change create any flow problems?
- Poor value: Could the benefits of this product or change be worth less than they cost?

In the Evaluation Phase you will find that there is plenty of cold-blooded judicial thinking and that this is an exercise in decision making. Decisions are based on experiments. The decision-making that takes place in this phase is that of making a choice between alternative methods of achieving a desired end. Thus, the Evaluation Phase involves the ANALYSIS of the alternative ideas generated during the Speculation Phase and then the DEVELOPMENT of the best-value alternative that meets the desired end.

ANALYSIS OF ALTERNATIVE IDEAS

If you have generated a sizeable number of ideas in the Speculation Phase, it would be wise to follow some sort of an elimination procedure, such as a three-round elimination tournament in order to help decide which ideas are to be developed.

Preliminary Screening

You will be basically rating or ranking the potential of each idea based on relative merits as Good, Fair, and Poor, as you would separate wheat from chaff. You do that by simply grouping the ideas into three basic categories:

Practical (Acceptable or Possible): These are the can-be-done ideas. This means that, at first glance, the ideas have real or possible merit, and thus should remain in contention.

Feasible (Maybe or Probable): These can be done, but there are questions that must be answered first. The idea might have merit and is still worth considering.

Impossible (Blue Sky or Not Now): These alternatives may be unreasonable for this time, but don't ignore such ideas. Be sure to save such ideas because what once was considered a blue-sky idea could turn out to be a viable idea.

In-Depth Analysis

This is the second round of the elimination tournament. It is the intermediate screening of the semifinalists. Here, a more analytical study takes place. In the Preliminary Screening round, decisions were more of an intuitive feel about the worth and potential of the ideas. Not so in this step, where experience, knowledge, and in-depth discussions are brought to bear.

Here, each Practical and Feasible idea is examined carefully. All Feasibles must be moved up to the Practicals category or down and out to the Impossibles. The Practical ideas need to be whittled down to a manageable number of potential winners. This is a tough task and takes lots of time, but the ideas that survive this round become the embryos of new and more efficient operating systems.

Evaluative criteria such as the following could be used to perform the in-depth analysis for the various alternative ideas being considered: reliability, quality, producibility, operability, maintainability, durability, safety, weight, appearance, human factors, logistics, packaging, schedule, interchangeability, cost savings, ease of implementation, and environmental factors. Still other evaluative criteria to be employed depending on the complexity or simplicity of the projects are force-field analysis, impact analysis, risk analysis, ROI analysis, cost-benefit analysis, break-even analysis, trade-off analysis, advantage-disadvantage scale, T-chart comparison of good and bad features, and others. These are some of the best tools in practice, but you will have to develop the evaluative guide that suits your needs.

Measurement Tools

Value Analysis	Benefit/Cost Analysis	Productivity Measurement
$\dfrac{\text{Function}}{\text{Cost}}$	$\dfrac{\text{Benefit}}{\text{Cost}}$	$\dfrac{\text{Output}}{\text{Input}}$

Customer Satisfaction Measures; Financial Measures;
Productivity Measures (Internal Measures: Product, Process, People)

FIGURE 10.1
Measurement tools.

Final Selection

This is the last round of the tournament, where the best ideas finally emerge as the first, second, and third choices. Here you identify the short list of ideas with the greatest potential to achieve value improvement. These final few candidates are evaluated by estimating their cost and evaluating their worth. Then those ideas that offer the greatest potential are selected for further development.

What is measured can be improved and the following measurement tools (Figure 10.1) could be helpful.

There is heavy use of nonfinancial measures in VA/VE, such as, part-per-million defects, percentage yields, scrap, unscheduled machine down-times, first-pass yields, and number of employee suggestions.

DEVELOPMENT

The objective of this step is to develop those one to three best ideas selected into practical value alternatives as a solution to the functional problem. This simple 2 × 2 matrix (Figure 10.2) helps present essential data visually in a meaningful way to assist the task of selecting the best ideas.

The modus operandi of this step is first to plan an investigation program and then execute the investigation, that is, develop the value alternative.

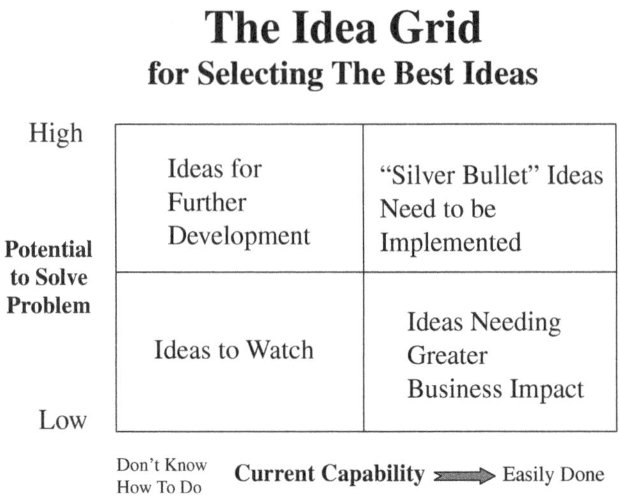

The Idea Grid
for Selecting The Best Ideas

High		
	Ideas for Further Development	"Silver Bullet" Ideas Need to be Implemented
Potential to Solve Problem	Ideas to Watch	Ideas Needing Greater Business Impact
Low		

Don't Know How To Do   **Current Capability** ⟹ Easily Done

FIGURE 10.2
The idea grid.

This is done by pursuing thoroughly each of the best ideas through *in-depth consultation with specialists and suppliers.* Consult specialists (outside and inside expertise), for they can help solve not only the performance problem but also the cost problem. Consult suppliers because they have people and facilities with special skills and knowledge. For consultation, we must provide complete information on the functional requirements so that those consulted can make detailed recommendations. From these recommendations, the final best-value alternative will be determined.

Refine ideas: rearrange, combine, modify, and develop. Considering what other questionable patent issues as well as other situations that might add costs is part of this step. Compute the total cost of the alternative, including cost of implementation. Verify the prototype.

At this point the core elements of the Information, Speculation, and Evaluation phases are completed and we now move on to the aftermath: the Recommendation, Implementation, and Audit phases.

11

The Recommendation Phase

CANDY WRAPPER CASE STUDY

After the ideas that have the greatest chance of success have been reviewed during the Evaluation Phase to make sure that the best value is being presented, the next step for the value analyst is to prepare a value-improvement proposal. In our candy wrapper case study, based on the 17% cost-reduction potential for the 50% inking, and 24% cost-reduction potential for the 25% inking offered by the current supplier, the value analyst recommended placement of trial orders in the amount of ten tons of each type.

If a value analyst or a project team can make the decision and implement the change, as would be the case with an executive or a self-directed cross-functional team, then the Recommendation Phase would be nonexistent. But if decisions by others are required, as often happens, the method of documentation in this phase is of great importance. However, one needs to be guided by the Paretonian statement that VA/VE is about 20% finding a better way and 80% selling it, meaning, it is more difficult to sell a creative idea than to create it! Hence, we need to benefit from the knowledge of the disciplines of sales and marketing.

OBJECTIVES OF THE RECOMMENDATION PHASE

There are three objectives of the Recommendation Phase:

1. To gain approval from others for the implementation of the proposed change, you need to prove it is better value. No amount of speech can make up for lack of substance. You need to know who has to be convinced. You want to win acceptance, but you should not plead for acceptance or implementation or approval, but be willing to settle

even for a test approval. You need to be flexible. If an idea is injected into your proposal, be willing to consider it. Don't win the battle and lose the war. It is also a good idea to have a backup proposal.

2. To document the project worker's or team's efforts, thus providing another example of success and serve as institutional memory.
3. To provide a training aid during the implementation of the proposed change.

PROPOSAL PREPARATION AND PRESENTATION

A value-improvement proposal form is provided in Appendix C to facilitate preparation of the presentation.

The proposal presentation step involves communication. According to leadership writer and speaker John C. Maxwell, "Everyone communicates, few connect." You need to make a direct connection with your readers or listeners. You don't want to be accused of favoring style over substance. Nor do you want to hear complaints that there is too much substance and not enough style in your presentation. The right feel and appearance of military equipment that contributes to soldiers' morale is a desirable form of esteem value. On the other hand, when form fails to follow function, the presentation is in trouble because of the false esteem value. ***You need to be passionate about an idea or product you are selling.***

Your value-improvement proposal would be either a verbal or written presentation. During a verbal presentation only key points should be highlighted and use of visual aids is highly recommended. If on the other hand, a written proposal is to be submitted, it should be tailor-made for the particular audience who has authority to accept the proposal, provide the needed resources, and approve implementation.

THE CONTENT OF THE VALUE-IMPROVEMENT PROPOSAL

The content of the proposal varies with the circumstances and the subject. But the following five parts normally represent the essential content of the proposal:*

---

* Carlos Fallon, *Value Analysis to Improve Productivity* (New York: Wiley-Interscience, 1971), p. 235.

1. **The Project**: Provide a brief description of the project studies with a summary of the problem: the item value-studied and what it was about, including identification such as name, part number, and so on.
2. **The Team**: The members, their position titles, and particular chore in the task group, such as team leader, recorder, and so on.
3. **Recommendations**: What are the *expected benefits*? What sorts of improvements, both quantitative and qualitative, will result from the proposal?
4. **Cost, Risk, and Net Gains**: What are the *unexpected expenditures*? Present this as a business venture, giving appraisal or risk, the payoff period, the breakeven point, and implementation cost. Explain advantages and disadvantages.
5. **Implementation Plan**: What are the *unexpected efforts*? What must be provided or accomplished? Who will approve it? Who will do it? Where will it be done? When will it be done? What are the target date and timetable for results? How will the funds be charged? What will be the schedule for reporting progress and follow-up?

THE A.I.D.A. MARKETING TECHNIQUE

Proposal presentation involves selling an idea and you have to *buy* your idea before you can *sell* it. You can employ the communicative and persuasive marketing and advertising technique known as A.I.D.A. to sell your proposal successfully. The term and approach are commonly attributed to American advertising and sales pioneer E. St. Elmo Lewis. The acronym stands for *attention, interest, desire,* and *action*.

Arouse *attention* (awareness) to the need for change. For example, a building will be cleared very promptly if someone reports that a bomb has been placed in it. To attract attention, you can say, "Here is how you can end your money worries!" Startle them with something unexpected. Give them important news. Try to express, not impress.

Create *interest* in change. You can arouse interest and hold their attention by quickly answering these questions that come to the minds of your audience. What is it? What will it do for me? Why should I bother about it? Why take time to consider it?

Develop *desire* for change. Once the audience's attention has been aroused and interest in seeking change created, then comes the time to develop a desire for the proposed change. To awaken desire to own, subscribe to, or do whatever the presentation suggests, enumerate the user benefits like improved delivery, improved performance, improved reliability, better quality, reduced weight, and better appearance. Reinforce your conclusions by benefit selling.

Gain a commitment to *action*. Your suggestion must impel action by a deadline or penalty. Express a sense of "the fierce urgency of now." What you want is positive action. Salesmanship skill in closing deals would be helpful. You've heard about the salesman who can sell refrigerators to the Eskimos!

Nothing happens until a sale is made!

GUIDELINES FOR JUSTIFYING THE NEED FOR CHANGE

- Will the change provide a better (faster, more current, more marketable, or more accurate) product or method? If so, how?
- What impact will the proposed change have on existing methods and systems?
- How much will the change cost in total expended dollars?
- How much will be saved in money or resources?
- How does the change fit in with future plans?
- Can the change or improvement be used elsewhere, and at what cost?
- Does the change provide for business fluctuations, both by expansion and contraction?
- What are the social, world-economic, and ecological implications of the proposed change?
- What will happen if the change isn't made?
- Why does the manager endorse the change, and why hasn't it been implemented earlier?

Build a better mousetrap, Emerson once wrote, and the world will beat a path to your door. But in today's market place, you need to sell your mousetrap by proving the need for it.

NEGOTIATION

In light of the fact that it is more difficult to sell a creative idea than to create it, it is essential for VA/VE to integrate the discipline of negotiation into the Recommendation Phase as that could help gain advantage in the outcome of the presentation. Effective negotiation between presenter(s) and the decision makers(s) can help in resolving conflicts and crafting outcomes that will satisfy various interested parties. It is good to keep in mind that the art of negotiation is to get people to do what you want them to do and make them believe it is their idea. It is also ancient wisdom that you don't win on the strength of your argument, but on the strength of your relationship with the person you want to persuade. Negotiations occur for two reasons: (1) *deal making:* to create something new that neither party could do on his or her own (interdependence), and (2) *dispute resolution:* to resolve a problem or dispute between the parties.

While there are very many books published on negotiation, the one that stands out among the best is *Getting to Yes: Negotiating Agreement without Giving In*, by Roger Fisher, William Ury, and Bruce Patton.* The two most common modes of negotiation are known as win-lose and win-win. "Win-Lose" refers to positional (bargaining), adversarial, competitive, fixed-pie, and distributive negotiation. "Win-Win" refers to

---

* Revised edition (New York: Penguin, 2011).

principled negotiation, joint problem solving, cooperative, variable-pie, and integrative negotiation. The main principles of these two modes of negotiation are shown here.

Positional Negotiation	Principled Negotiation
Open high or low	Use objective standards
Trade concessions toward midpoint: compromise difference	Choose from many options rather than splitting the difference
Disguise true feelings: wear a mask	Speak openly and clearly, describing your interests
Discredit case and claims made by the other party	Accept case made by the other party as one possible solution
Use tactics to keep the other party off balance and feeling threatened	Make sure the other negotiator feels comfortable, secure and respected

We can view the general framework of negotiation as a three-stage process consisting of

1. Prenegotiation
- Who is to negotiate?
- Where will the negotiations be held?
- Gather intelligence
- Determine objectives
- Recognize your strengths and weaknesses

2. The Actual Negotiation
- Use appropriate techniques including:
  - A properly framed agenda
  - Careful questioning
  - Trade concessions
- Know your BATNA (Best Alternative To a Negotiated Agreement)
- Watch out for the winner's curse (Unpleasantness on your part will probably bring a similar behavior from the other side.)
- Weigh up counterparts and use appropriate behavior
- Recognize ploys

3. Postnegotiation
- Detail the agreement and circulate it among stakeholders
- Sell the agreement
- Implement and monitor

Negotiation is both art and science. The art in negotiation is concerned with the human dynamics, the human side of negotiation, relationship building, and the soft skills. The science part deals with accurate observation, realistic assumptions, correct factual analysis, logical inferences, planned behavior, and optimal presentation for each moment of a changing bargaining situation. Although an experienced negotiator will use a high degree of skill and creativity (the art), the distinction is that, as an art, there may be several different approaches from different points of view that will each achieve the desired objective. In scientific negotiation, there are fewer best ways (strategy and tactics) of proceeding at any given moment to achieve the desired objective. Congruent with the VA/VE methodology, and viewing negotiation as a joint problem-solving approach, the following five steps will help facilitate the work of the Recommendation Phase.

The Five Stages of Negotiation

1. **Creating the Climate**
   - What ground rules and agenda will I suggest?
   - Will I go first or will I let them present their position first?
   - What will I do to create an open climate?
2. **Defining the Problem**
   - How will I state my position clearly and firmly?
   - How will I make sure that I understand and clarify the other party's position?
3. **Understanding the Problem Fully**
   - Have I made a list of my best guess about the interests and needs behind the other party's position?
   - Have I done the same for my own position?
4. **Generating Alternative Solutions**
   - Have I made a list of some creative options for the other party?
   - Have I done the same for myself?
5. **Evaluating & Selecting Alternatives**
   - How will I make sure I make it a package deal?

As Dr. Chester Karrass reminds us, "In business, you don't get what you deserve, you get what you negotiate." I might also add that when you combine your negotiation skills with the value methodology, you may turn out to be a black-belt negotiator.

RECOMMENDATION PHASE CHECKLIST

- Did you select the right spokesperson to present your value-improvement proposal?
- Did you have a back-up proposal?
- Did you use good graphic illustrations?
- Did you show flexibility and a collaborative demeanor?
- Did you use the manager's language, like "performance improvement" and "cost-effectiveness"?
- Did you craft an implementation plan?
- Did you make a compelling case to get positive feedback?
- Did you just communicate or did you connect with the needs of the customer?
- Did you chair-fly or rehearse your presentation?
- Did you share credit with all team members?
- Did you evoke "the fierce urgency of now" and impel action?

Nothing that is worthwhile is very easy. Remember that.

—**Nicholas Sparks**

DECISION REQUIREMENTS OF THE RECOMMENDATION PHASE

There are two decision requirements of the Recommendation Phase: Improvement or change authorization and expenditure authorization. Please refer to Appendices C and D for guidelines.

12

The Implementation Phase

CANDY WRAPPER CASE STUDY

The objective of the Implementation Phase is to ensure that the proposals that were approved under the Recommendation Phase are translated into action in a timely manner in order for the organization to achieve the expected results. Applying VA/VE to produce creative solutions to problems is the beginning of the change-management process, and management then needs to execute the plan to achieve its objectives.

In the Recommendation Phase of the Candy Wrapper case study, the value worker recommended that the company order 10 tons each of candy wrappers with 25% and 50% inking based on cost-reduction estimates. Upon receipt of the trial shipments, a market survey was conducted to determine customer reaction to the design changes of the wrapper. With customer satisfaction confirmed, management approved placement of annual orders of the 50% inking for years two and three. In the middle of the fourth year, the 50% inking was discontinued and replaced by the 25% inking.

CREATIVITY IS NOT ENOUGH!

A powerful new idea (like VA/VE itself) could be kicked around unused in a company for years, not because its merits are not recognized, but because nobody has taken the responsibility for converting it from words into action. What is often lacking is not creativity in the idea-generating sense, but innovation in the action-producing sense, that is, putting ideas to work.

Following Schumpeter (1934), contributors to the scholarly literature on innovation typically distinguish between *invention*, an idea made manifest, and *innovation*, ideas applied successfully in practice. In economics, the change must increase value, either customer value or producer value. Those who are directly responsible for the application of inventions are often called pioneers in their field, whether they are individuals or organizations—hence, the distinction between paradigm shifters (creativity) and paradigm pioneers (innovation).

The purpose of organizations is to achieve the kind of order and conformity necessary to do a particular job. But creativity and innovation disturb that order. Hence, organizations tend to be inhospitable to creativity and innovation, but without them organizations would eventually perish. Creativity and innovation are the engines that drive lasting success and help you leapfrog the competition rather than constantly playing catch-up. Creativity is the beginning, the spark. It is the ideation process, and innovation is the implementation of ideas. That makes VA and *innoVAtion* go perfectly together.

Implementing a proposal that results in a modest saving is far more important than publishing proposed dramatic savings that are never achieved. While we will be discussing a variety of methods of organizing a value program in Section III, it is important to mention here that the value team's approach is the most advantageous as it increases the probability of implementation, especially if the team members have sufficient management stature to approve implementation in their respective departments.

SETTING UP AN IMPLEMENTATION PLAN

Given here is a checklist of principles and steps for setting up an Implementation Plan:

- Base your plan on the need for speed in implementation—the bias for action.
- Translate ideas into action.
- Set up specific milestones for each event that must be executed to carry out the decision.
- Provide a road map for who does what and when.

- Set up a warning system that flags potential problem areas as early as possible.
- Obtain skills, facilities, and other resources.
- Develop measures and measurement systems.
- Pilot, improve, and roll out!

IMPLEMENTATION BARRIERS

The implementation of proposed changes can be jeopardized by neglecting to anticipate roadblocks such as financial concerns, corporate culture, lack of management support, lack of expertise, and unclear direction. Some more barriers are listed here:

- Senior management does not recognize the value that the value-improvement effort will generate.
- Insufficient mandate to facilitate and drive change.
- Cultural and organizational barriers:
  - Highly fragmented organizations that limit coordination and exchange of information and best practices.
  - Purchasing or procurement viewed as a "backwater" function.
  - Nontraditional purchases viewed as being best managed by end-users.
- Lack of technical, specialized, or general business skill sets.
- Inadequate information to enable easy, value-added analysis.
- Continued focus on headcount reduction.
- Risk taking is not rewarded.

IMPLEMENTATION ENABLERS

On the other hand, the prospects of ideas coming to fruition are improved where enablers like the following are nurtured:

- Benchmarking with best-in-class organizations serves the role of enabler because of legitimacy.
- Open and obvious recognition from senior management supported through ongoing communication, active involvement, and attractive incentives.

- Complete user-friendly access to information on all purchases to drive value maximization.
- Extensive use of cross-functional teams to expand the quality of inputs.
- Performance measurement systems that track results.
- End-user needs and concerns are captured through formal and informal survey techniques.
- Consider piloting the solution.

REASONS FOR CONDUCTING A PILOT PROGRAM

A pilot is the trial implementation of the proposed and validated solution on a reduced scale. The purposes of a pilot program include to

- Minimize the risk of identifying potential problems
- Increase organizational buy-in
- Validate and refine cost and benefit estimates
- Perform adjustments and observe the solution implications
- Evaluate the effectiveness of measures used to monitor the improvement
- Test the validity of the solution
- Make mistakes on a small scale

13

The Audit Results Phase

CANDY WRAPPER CASE STUDY

During the second through the fourth year (while the original design was still in supply) actual cost savings realized on 221 tons of the new design wrapper (49 tons with 50% inking and 172 tons with 25% inking) amounted to $264,346.00.

The cost savings on the new annual requirement of 175 tons for the fourth year was as follows:

90% inking	$1,192,768.00
25% inking	− $963,477.00
TOTAL SAVINGS	$229,291.00

Based on those results, the company explored changing the supplier for further cost savings. The savings on 175 tons of candy wrappers was calculated as follows:

Current European Supplier	$963,477.00
New Korean Supplier	− $549,477.00
ANNUAL SAVINGS	$414,000.00

Those savings prove the saying that "Sweet comes from sweat"!

The standard management process is comprised of PLANNING to achieve predetermined objectives, ORGANIZING to put the plan into action, LEADING people to execute the plan, and MEASURING performance to assure conformance to the plan. We are now here at the end of the VA/VE process—the phase to measure and audit the results.

The operational excellence office or the project management office is responsible for the execution of this phase.

OBJECTIVES OF THE AUDIT RESULTS PHASE

There are many objectives in this important final phase:

- To measure and compare results to the desired state.
- To check for new problems and continuously improve.
- To monitor progress and keep projects on track.
- To follow up on ideas to their realization.
- To institutionalize best practice.
- To give successful VA/VE examples so employees can see how to do it.
- To document lessons learned and record VIP (Value Improvement Program) contributions.
- To create value-sensitive attitudes among employees and to develop a team spirit in the approach to VA/VE.
- To recognize value-improvement suggestors and implementers. This is probably the strongest selling point that can be made as sharing credit and taking responsibility for mistakes contributes to the success of the value improvement program.
- To celebrate success, because excellence deserves recognition.
- To formally close the project.

REVIEW AND REPORT

The last phase of a project should be a process review so that the management of projects can be improved. However, it should not be conducted only at the end of a project, but should be done at tollgates or major milestones in each phase of the Value Improvement Process. This also helps with the decision to continue or terminate a project. Periodic project-process review should be conducted in a spirit of learning rather than in a climate of blame and punishment.

A review should be followed by a report detailing planned versus actual results and comparing progress to the original plan so that action can be taken to correct for deviation from the plan. This phase involves the frequent review and removal of barriers. Here you check for real business impact by performing a **value-for-money audit**. You continuously communicate progress to executive leadership and those involved in projects.

The content of a report and extent of comprehensiveness will depend on the method of reporting or communication selected (newsletter, house organ, corporate website, or lobby display) and the audience to whom the report needs to be addressed. Whichever the case, such progress reports commonly include these elements:

- Number of VA/VE projects currently under study
- Estimated potential savings in items under study
- Number of Value Engineering Change Proposals (VECPs) currently under evaluation, either in-house or by suppliers
- Breakdown of the age of the proposal under evaluation: 0–90, 90–180, over 180 days

MONITORING AND EVALUATION

Monitoring and Evaluation focuses on efficiency, effectiveness, and impact of the project, that is, the value study. Efficiency tells you that the input into the work is appropriate in terms of the output. This could be input of resources utilized, such as money, time, labor, equipment, and so on. Effectiveness is a measure of the extent to which a development program or project achieves its specific objectives. A synonym for effectiveness is efficacy meaning the ability to produce a desired or intended result. Impact tells you whether or not what you did made a difference to the problem you were trying to address. Monitoring & Evaluation in the case of development work enables you to check the bottom line—not "Are we making a profit?" but "Are we making a difference?" It would be worth mentioning here that the late Peter Drucker taught us that efficiency is doing a thing right and effectiveness is doing the right thing.

EARNED-VALUE ANALYSIS

Earned-Value Analysis compares the PLANNED amount of work with what has actually been COMPLETED, to determine if COST, SCHEDULE, and WORK ACCOMPLISHED are progressing as planned.

Have we done what we said we'd do? This is analyzed in terms of the percentage of complete estimating: budget spent, work done, and time elapsed. Work is "earned" or credited as it is completed.

INSTITUTIONALIZATION OF CHANGE

Once the change has been determined to be successful, it is integrated into business operations in these ways:

- Standardize best-in-class management-system practices.
- Diffuse best practices in the organization. Without organization-wide application of best practices, value-improvement objectives cannot be fully achieved.
- Integrate standardized best practices into policies and procedures.
- Transform how day-to-day business is conducted. Far more than completing projects, the objective is transformation of culture.

RETURN ON INVESTMENT

The Audit Results Phase involves monitoring, measuring, and reporting to management the savings-to-cost ratio of the value-improvement program. In short, management will be able to know through such a reporting system the return on investment from its value program.

The U.S. Army Corps of Engineers (USACE) value engineering/value management program has been a leader in applying the Value Engineering Methodology to construction projects since 1964, solidly demonstrating the Corps' cost effectiveness. Historically the program has returned 20 dollars for each dollar spent on the VE effort, and has resulted in construction of over $3.8 billion in facilities, without additional funds requests to Congress.

The following are net USACE VE/VM savings and cost avoidance for the five fiscal years 1998–2002 as reported to the Department of Army and Office of Management & Budget (OMB).* The figures do *not* include hundreds of millions of dollars in cost avoidance from work performed for other agencies or countries, or from innovations subsequently repeated in similar projects.

---

* Value Engineering/Value Management Headquarters, US Army Corps of Engineers (CECW-EV).

Year	Military	Civil Works	Total Savings
FY 98	$63,751,000	$86,091,000	$149,842,000
FY 99	$56,223,000	$103,742,000	$159,965,000
FY 00	$45,109,000	$80,251,000	$125,360,000
FY 01	$38,780,000	$90,777,000	$129,557,000
FY 02	$95,764,000	$96,568,000	$192,332,000

AUDIT RESULTS PHASE CHECKLIST

- Did the idea work?
- Was money saved?
- Was the design improved?
- Can it benefit others?
- Has the change had proper publicity and distribution?
- Should any awards be made?

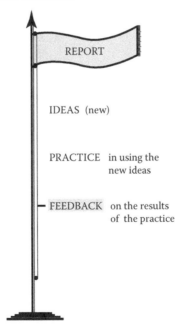

OPERATIONAL EXCELLENCE is demonstrated by results!

1. PROBLEM
2. SOLUTION
3. RESULTS

Section III

The Value Organization

It is imperative for corporations to put the same energy used for new products and processes into organizational design. With the right design, your organization will have the capabilities to pursue whatever strategy is necessary to compete on any scale, react to any market change, leverage any opportunity, and sail past the competition.

—**Lowell L. Bryan and Claudia I. Joyce,** *Mobilizing Minds: Creating Wealth from Talent in the 21st-Century Organization*

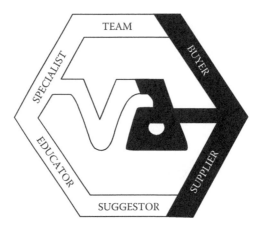

Now that you have read the Introduction, and Sections I and II of this book, and appreciate the critical relevance and significance of Value Analysis/ Value Engineering, the next step is to learn the alternative organizational approaches to start a value program or an operational-excellence program in your company. Depending on the size of your company or the type of industry you are in, you will set up your value program in one or a combination of the following organizational approaches.

14

The Value Specialist Approach

It has often been said that "a specialist is a person who knows more and more about less and less, until (s)he knows almost everything about nothing. A generalist is a person who knows less and less about more and more, until (s)he knows practically nothing about everything."

A value specialist is a full-time value analyst or value engineer trained in the VA/VE discipline of obtaining essential function at least cost in resources without impairing quality. Value specialists may or may not be holders of a Certified Value Specialist (CVS) certification, though it definitely helps to have one. They are the persons in charge of the corporate value-services function, also known by other names like operational-excellence center or project-management office.

Here are the attributes of value specialists:

- They wear a Sherlock Holmes hat, meaning that they have a detective's mind and would be able to detect problems or opportunities.
- Their eyes have dollar signs in them, meaning that they have cost sensitivity or cost visibility or cost discipline.
- They have the audacity and claim the right to challenge specifications, while the user department maintains the right to change specifications.
- They don't accept the notion that "it can't be done!"

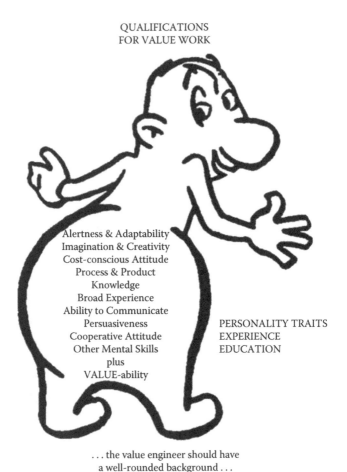

QUALIFICATIONS
FOR VALUE WORK

Alertness & Adaptability
Imagination & Creativity
Cost-conscious Attitude
Process & Product
Knowledge
Broad Experience
Ability to Communicate
Persuasiveness
Cooperative Attitude
Other Mental Skills
plus
VALUE-ability

PERSONALITY TRAITS
EXPERIENCE
EDUCATION

. . . the value engineer should have
a well-rounded background . . .

The value specialist must be a jack-of-all-trades and master of value.

If you wish to have a successful value program or operational-excellence program, you need to have someone responsible for it, because when everyone is responsible no one is responsible. The value program or operational-excellence program is like a wheelbarrow: it will stand still unless somebody pushes it, hence the need for the full-time value specialist.

One organizational approach common in progressive companies is to have a corporate value-improvement staff reporting to an executive with the power to cut across departmental or divisional lines. A more common practice, however, is to give VA/VE a residence of convenience in the department having the greatest need and potential for performance and cost improvement, such as engineering, manufacturing, or purchasing and materials management. Where there is responsibility for expenditure exceeding 53 cents of every sales dollar, there could be a compelling argument to have an active VA/VE activity in a procurement or supply management organization.

The success of value specialists will depend on these factors:

- Management's support for their activities.
- How well they are situated within the organization so that they can perform their duties properly.
- How well qualified they are by training, experience, and personality to carry out the assignment. They must gain technical competence by visiting suppliers, reading up-to-date professional journals, and attending seminars, conferences, and tradeshows.

The job title of the value specialist could vary from Value-Program Administrator to Manager of Value Analysis, Director of Value Services, or Vice President of Operational Excellence. Whatever the title, the job of the value specialist may be classified as a coordinating and an operating function. In smaller organizations these two functions may be performed by the same VA/VE staff, yet they remain separate and distinct.

The coordinating function is concerned with overall program planning and control, determination of priorities, assignment of targets and allocation of resources necessary to meet these targets, measurement of progress toward targets, development and supervision of the VA/VE training program, design of the publicity program, and development of policy and procedures for the efficient and effective management of the value program. Keep in mind that the goal of VA/VE is commitment to initiating value studies to achieve value improvement and not necessarily the traditional cost-reduction targets.

While the coordinating function is primarily characterized by its assistance to those who perform the value studies, the operating function is concerned with the actual performance of the value studies and submission of Value Engineering Change Proposals (VECPs).

The **coordinating functions** include:

INTEGRATION
- Introduce the value of VA/VE to all stakeholders.
- Promote efficient and collaborative teamwork among all functions.
- Integrate and liaise other value activities in other functional areas.

- Make value sensitivity contagious in the organization.
- Deploy VA/VE as the integrative methodology.

EDUCATION

- Involve formal and informal training of personnel to (a) improve their value ability and cost consciousness, (b) develop creativity, (c) promote a constructive discontent, and (d) demonstrate proven techniques by which the desired function can be performed at less cost.
- The regular value training must be augmented with and strengthened by examples and workshops to highlight the opportunities and methods for lower costs and service improvement and publicly display before and after case studies. Efforts must be made to maintain institutional memory.

The **operating functions** include:

CONSULTATION

- A staff position free from line responsibilities, the Value Specialist is concerned with the state of the art and attends seminars, symposiums, shows, and the like to keep abreast of new developments in materials, machines, and standards, and visits suppliers to serve as internal management consultant on operational excellence.
- Serve as a catalyst for change and transformational management.

EVALUATION

- Assist project teams in product and process functional analysis and total cost analysis.
- Assist line mangers in the development of value proposals for decision and implementation by user departments.

Hence, organized systematically and staffed competently, a value-services function or office of operational excellence can contribute tremendously for the success of an organization, keeping in mind that distinctive capabilities are essential to distinctive performance.

15

The Value Team Approach

A team is a small number of people with complementary skills who are committed to a common purpose, performance goals, and approach for which they hold themselves mutually accountable.

—**Jon R. Katzenback and Douglas K. Smith,** *The Wisdom of Teams: Creating the High-Performance Organization*

The value team approach is what is commonly referred to as the task force, project team or cross-functional sourcing team. Although a value study can be accomplished normally through an individual effort, performance improvement and cost-effectiveness could be enhanced when the team blends their talents toward a common objective. Value-minded culture will bind your team together into a whole much bigger than the sum of its parts.

The value team is usually comprised of five to eight people with diverse backgrounds relevant to the specific study. Differences are sources of strength—and perhaps the most underrated asset in management. Without differences, creative relationships would be impossible. For best results, the team members should receive VA/VE training prior to participation in the value study.

FROM BUREAUCRACY TO ADHOCRACY

In the world of bureaucracy where change meanders rather than leaps, we need to reinvent that pyramidal architecture of "command-and-control" and replace it with the new culture of teamwork that enhances adaptability and innovation. Under bureaucracy, specialization separates people.

With the team approach, VA/VE brings them together. VA/VE provides a method for integrating the specialized skills of engineering, purchasing, accounting, marketing, and other functions in order to make sure that the entire organization contributes to making the right product for today, instead of making yesterday's product right.

There is no limit to what ordinary people can do when they combine their skills to support each other in pursuit of a common goal. In a soccer game, all members of the team depend jointly on each other. Each is a specialist contributing to the success of the team. Similarly, with the value team approach, everybody contributes to the value improvement. The value of the value team is in its being synergistic, that is, the total effect of combining everyone's knowledge and experience will be greater than the sum of its parts. Hence, we have learned that TEAM stands for: *Together Everyone Achieves More!*

HOW TO ORGANIZE A VALUE TEAM

Since the value team is the most popular organizational approach in use, let's look at how you would go about organizing the team, in terms of team composition and other factors.

Goals of Team Organization

- The necessary combination of skills.
- Relevant and contrasting knowledge for the project at hand.
- Depending on the nature of the value study, ensure the necessary balanced combination of skills to consist of marketing, engineering, purchasing, production, and accounting (cost-estimating).
- Objectivity through compensating biases.
- Representation of departmental interests (people with authority).

Steps for Organizing Project Teams

- Establish good human relations from the beginning. Because VA/VE is concerned with creating change, the concern is with human relations. In VA/VE, there is a high degree of dependence on cooperation with other people. Therefore, good or poor human relations can relate directly to success or failure of the project:

- Appoint a team leader, secretary (recorder), and advisor.
- Ensure interest in as well as knowledge of VA/VE.
- Identify appropriate team members.
- Clarify responsibilities.
- Determine objectives and goals.
- Develop a project plan.
- Define team success.
- Provide necessary value-training.

STAGES OF TEAM DEVELOPMENT

The business world doesn't function top-down as in the military. Today's leadership needs to use persuasion and influence and promote teamwork. The following stages of team development explain how people adapt to teamwork:*

Forming Stage: People are concerned with how they will fit in and who makes decisions. A leader would provide structure. Otherwise, informal leadership emerges.

Storming Stage: This is frustrating to most people as people begin to question their goals, roles, and procedures.

Norming Stage: At this stage people are beginning to resolve their conflicts and take responsibility for the team's goals, procedures, and behavior.

Performing Stage: This is when people tend to produce results and the leader's job becomes easier. At this point they are a TEAM.

The style of leadership appropriate for a team depends on its stages of development:

- In the forming stage, it is directive.
- In storming, it is influencing.
- At the norming stage, the team leader switches to a participative style.
- Finally, when the team reaches the performing stage, you can be delegative.

* James P. Lewis, *Fundamentals of Project Management* (New York: AMACOM, 2002), pp. 123–124. Identified by Bruce W. Tuckman (*The Six Sigma Handbook*, p. 71).

It took **teams** to build them!

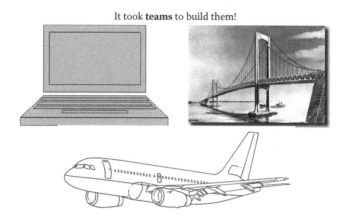

"When was ever honey made with one bee in the hive?"—Thomas Hood, 1826

FIGURE 15.1
The power of teamwork.

Figure 15.1 illustrates the power of teamwork and confirms that "None of us is as smart as all of us," as Ken Blanchard has said. Individuals didn't make it to the moon—NASA did. Teamwork or team spirit is the fuel that allows common people to attain uncommon results. The ultimate champion is the team.

Your vast procurement expenditures require Value-Adding Sourcing Teams (VAST) equipped with VA/VE to achieve operational excellence. As the African proverb teaches us, "If you want to go fast, go alone; if you want to go far, go together."

16

The Value Buyer Approach

In the performance of any buying function—from shopping for groceries to procurement for a large company—the consideration of "getting the best for the money spent" is ever present. Before the term *Value Analysis* came into being, buyers might have effected a considerable saving in the purchase of some item wherein they may have used the VA methodology without being conscious of it. The savings were taken for granted; since this was the job they were expected to do. The successful application of formal VA/VE demonstrates the buyer's level of professional competence. However, one of the stickiest questions in purchasing is, when can a buyer justifiably claim credit for having saved the company money?

Legitimate cost saving is a proved saving realized through the application of the buyer's ingenuity, imagination, judgment, objectivity, and research over and beyond the efficient performance of normal responsibilities of purchasing. For instance, if a buyer suggests specification changes relating to material substitution or component redesign, the buyer is obviously entitled to credit for any savings that result. It is equally apparent on the other hand, that no buyer deserves credit for savings that come from industry-wide price decreases or suppliers' voluntary price concessions and so on.

VA/VE has elevated the professional status of the purchaser from a reactive, tactical responsibility to a proactive, strategic player and a bottom-line contributor. VA/VE enriches the buyer's job. But one basic obstacle to individual buyers being their own value analyst is lack of time. The individual buyer has requisitions to handle, inquiries to make, bids to review, salespeople to interview, orders to place, and all kinds of other duties to perform. But VA/VE is not an additional chore tacked onto the multitude of tasks that each buyer performs in a working day. Rather, it has always been an integral part of a value-minded buyer's responsibilities. In fact, buyers are in a much better position to practice VA/VE because

- VA/VE is function oriented and every purchase order buys a function.
- Buyers are familiar with the requisitions and the functions of the items they buy for the various user departments or internal business partners.
- The buyer is the manager of external production and is the one who has the mandate to deal with suppliers.
- The nature of the buyer's work is analytical. Every good purchase is the result of a study and evaluation of requirements, a search for alternative materials and alternative sources of supply, in an effort to get the best return on the procurement dollar spent.
- Buyers are professional customers. They are commercial intelligence agents. They are commodity specialists.

Hence, the buyer's duty is to challenge wasteful and avoidable costs inherent in the goods and services that need to be purchased. The buyer's job is to add value!

Conventional purchasing, that is, buying exactly what is requisitioned without functional analysis, is activity oriented. But value buying is result oriented. Table 16.1 reviews the profitability of value buying. It shows that a company's profit increased 10% from Year 1 to Year 2 due to an increase in sales. Note however that all production costs also increased, limiting the increased profit to just 10%. Value Analysis reveals that if sales remained the same but material costs were reduced by just 10%, overall profit would increase by 45%! That is a very large result for a small change in value buying.

TABLE 16.1

The Value Buyer Approach

	Year 1	Year 2	Reduced Material Cost 10%
Sales	1,000,000	1,100,000	1,000,000
Material	450,000	495,000	405,000
Labor	200,000	220,000	200,000
Overhead	250,000	275,000	250,000
Total cost	900,000	990,000	855,000
Profit	$100,000	$110,000	$145,000
Increased profit		+10%	+45%

From the foregoing example, you can appreciate that purchasing is about saving, not spending!

The overarching goals of public procurement policy are to

- Assure the best value for taxpayers' money by professionalizing the procurement staff
- Develop responsive and responsible suppliers or vendors
- Achieve efficient and effective contracting processes to assure *value-for-money* from suppliers

Please refer to the various best-practice examples in this book (Under the Introduction and Chapter 13) cited for the contributions of value-buying in government operations.

When the goal is boosting profits by dramatically lowering costs, a business should look first to what it buys! The impact of procurement professionals on the quality, cost, and productivity of the organization is one of the keys to competitiveness in the global marketplace. For each dollar saved in procurement cost, approximately $0.60–$0.75 is carried to net profit. By contrast, for each dollar of incremental sales, profit increases by just $0.03–$0.05. This power of the purchasing function is summed up by Zeger Degraeve and Filip Roodhooft: "Purchased products and services account for more than 60% of the average company's total costs. For steel companies, that number goes up to 75%; it's 90% in the petrochemical industry. Even at service companies, the figure is typically a hefty 35%. Bringing down procurement costs can have a dramatic effect on the bottom line—a 5% cut in material cost can translate into a 30% jump in profits."*

Hence historically, purchasing that used to be a corporate backwater became a fast-track job as purchasers (value buyers) showed they can add millions to the bottom line with the practice of value analysis/value engineering. Furthermore, organizations now recognize that the biggest challenge in selling is smart buying.

---

* A smarter way to buy. *Harvard Business Review*, June 2001.

17

The Value Supplier Approach

The key to managing a business successfully is that you have a strong customer base and a stronger supplier base. So how do you decide if a supplier is good and trustworthy? The **ideal supplier** would be one who meets the following standards:

- Delivers material that is defect free and needs no inspection
- Delivers on time
- Makes frequent deliveries without penalties
- Requires a minimum of paperwork
- Responds quickly to product or information needs
- Is agile and flexible in responding to changing business conditions
- Offers technical assistance and ideas to improve products
- Charges a fair and reasonable price

It is important to make use of the knowledge of suppliers. They are sources not only of goods and services but also of ideas and technology, savings and profits, stability and growth, and competitive advantage.

A supplier's value-improvement proposals may be enlisted to mutual advantage. It is one of those happy situations in which everybody wins. Here is a sample of Allis-Chalmers Value Engineering Change Proposals (VECP) questionnaire from a form they use to solicit supplier participation in VA/VE program:

- Would you be interested in negotiating an incentive-sharing plan for your Value Engineering Change Proposal (VECP)?
- What is the function that is performed by this product?
- Could costs be reduced by relaxing requirements as to tolerance? Finishes? Testing? By how much?

- Could costs be reduced through changes in: Material specified? Ordering quantity? Supplier of raw material changed? Manufacturing process used, i.e., casting, forging, stamping, etc? By how much?
- Can you suggest any other changes that would: Reduce weight? Simplify the part? Reduce overall cost? By how much?
- Does it appear that any of the specifications or quality-control requirements are too stringent?
- In supplying this product, what is the greatest element of your cost that we might possibly help alleviate?
- Do you have a standard item that could be substituted for this part satisfactorily?
- Other suggestions?

In awarding contracts to suppliers, we need to keep in mind that contracting is about risk management. How much risk do we want a supplier to take, and how much do we want to maintain ourselves? If we have a **firm, fixed-price contract** then we have transferred the predominant portion of the risk to the supplier. The other end of the spectrum would be a **cost-plus contract**, where we have really assumed almost all of the risk ourselves. Here are the standard types of contracts:

- Fixed-price.
- Firm fixed-price: specific price throughout term.
- Fixed-price with adjustment: provides for price adjustment to offset inflation/deflation or changes in material or labor costs.
- Fixed-price with redetermination: price becomes fixed after future costs and amounts of labor or materials become known.
- Fixed-price with incentive: cost reductions are shared by buyer and seller; incentive for meeting or exceeding schedule targets.
- Lump-sum: specifies the total amount that the purchaser will pay for the completed job.
- Cost reimbursable.
- Cost plus fixed fee: seller is paid for allowable costs plus a negotiated fixed fee.
- Cost plus percentage of cost: rewards seller for inefficiency. Agreement covers all costs plus a percentage of added costs.
- Cost plus incentive fee: supplier shares savings if costs are below target.

- Cost plus award fee: seller's receipt of agreed fee is dependent on buyer's evaluation of its performance.
- Time and material contracts: fixed labor rate plus materials.

Suggestions from suppliers can play an important part in making your value program a success. Tables 17.1 and 17.2 provide an example on how an incentive-sharing plan could work to your mutual advantage.

TABLE 17.1

The Value Supplier Approach

	Sales Income	Profit
Original contract	$1,000,000	
Profit (6%)		$60,000
Gross VA savings	$120,000	
Implementation cost	$20,000	
Net VA savings	$100,000	
Revised (reduced) contract	$900,000	
Profit (6%)		$54,000
50/50 Sharing plan		
Customer	$50,000	
Supplier		$50,000
New profit to supplier		$104,000
Therefore, supplier profit increased by ———%		

TABLE 17.2

VECP Incentive Clauses

Incentive Sharing	Time Period	Percent Sharing
Instant contract	Duration of contract	50% sharing on net savings of current contract. Possibly 75%.
Royalty payments	1 year min.	Lump sum or 20%
	3 years max.	Alternatively,
		40% for 1 year
		30% for 2 years or
		20% for 3 years
Collateral savings	One typical year's use	10%–20% share of an average one year's life cycle ownership savings

18

The Value Suggestor Approach

If we agree that people are the greatest asset of an organization, then there is great need for judicious use of this asset. For management to learn through personal experience is expensive. When you receive suggestions from employees, you will be better than all of your years of experience. Besides, what better way is there to recognize the importance of your fellow employees than to ask them for their valued ideas? Shifting from the viewpoint of the all-knowing managerial decision makers to making more employees part of the corporate-strategy process is a powerful means of aligning them more closely with the company's overall direction and goals.

The employee-suggestion system gives people the opportunity to acquire the great pleasure of discovery—to create new ideas. People spend most of their waking hours at work. The late Steve Jobs tells us, "Your work is going to fill a large part of your life, and the only way to be truly satisfied is to do what you believe is great work; and the only way to do great work is to love what you do." If you are good at what you do and like your job, you are bursting with mojo.

Different companies have different suggestion-system policies, but improperly conducted suggestion systems are worse than none. Suggestion systems must primarily be sold as employee motivators, and secondly as a source of profitable ideas. With VA/VE providing the goal-setting and problem-solving ability and a suggestion system stimulating employee motivation, such intervention and empowerment can help organizations to effectively integrate employee creativity and management innovation. That is why innovative organizations recognize the great value of the entrepreneurial perspective of VA/VE and support the suggestion system's intrapreneurial movement.

I referred previously to Jim Collins's statement about meaningful work, that meaningful work is necessary for a meaningful and great life. I hereby confirm to you that the application of VA/VE makes work meaningful by providing people with the opportunity for self-actualization and achievement. Sustainable happiness comes from doing *Lots of Value Engineering*. Happiness drives higher workplace productivity, and it is the ultimate productivity booster! Happy employees make happy customers, and happy customers become loyal customers!

It is said that there are three kinds of workers: those who MAKE things happen; those who WATCH things happen; and those who WONDER what happened. When empowered with VA/VE and the employee suggestion system you can make things happen and thereby enhance employee satisfaction.

Having a suggestion system is management through the collective knowledge of all employees. In the United States, people who realized the value of this idea founded the National Association of Suggestion Systems in 1942 (now renamed Employee Involvement Association). The objectives of the employee suggestion system are to

- Provide a two-way communication system for employees and management
- Encourage employee creativity and management innovation
- Provide a natural medium for training in VA/VE and other improvement methods
- Stimulate employee identification with corporate objectives
- Emphasize the importance of individual accomplishment
- Unleash the full potential of employees
- Focus attention on the importance of employee ideas for organizational progress
- Boost employee morale and productivity through participation
- Recognize employees' constructive and creative contributions
- Reward employees for useful suggestions
- Provide means of increasing employee earnings
- Improve worker well-being
- Identify individuals with greater potential
- Bring to management's attention problems that otherwise might remain unidentified
- Develop a spirit of teamwork
- Improve quality and productivity
- Identify and eliminate waste
- Improve costs and increase profits
- Improve safety and working conditions
- Improve service and public relations
- Improve the corporate competitive position

The competitive world of business and industry takes full advantage of the employee-suggestion system; and governments that appreciate the value of employee involvement have also instituted such policies. For example, the Government Employee's Incentive Awards Act, Title III of Public law 763, enacted September 1, 1954, and established the Suggestion Program. The foundation of the Federal Suggestion Program is found in Title 5, Code of Federal Regulation, Part 451. The law established a Government-wide program encouraging all employees to improve the efficiency and economy of Government operations. The Federal Guide for the Employee Suggestion Program reads as follows:

The Federal Government benefits from implementing employees' ideas and suggestions. In order to maximize these benefits, all employees must become involved in the process of continuous improvement. The National Performance Review brings new emphasis to employee involvement and empowerment for improving Federal services.

Employees are the most likely source for workable ideas and suggestions on cutting red tape, putting customers first, and getting back to basics. Employee suggestions award/benefit ratio followed the typical ratio of 30:1, i.e., it receives approximately $30 in benefits for every $1 the Government spends on suggestion awards. Historically, about 25% of suggestions are adopted and rewarded. Ultimately, the Government continues to benefit from many suggestions for several years.

The worth of all the excellent suggestions approved cannot be measured only in dollars. Suggestions with intangible benefits also bring improvements in public services, including scientific and medical advances, and national security. Benefits from these adopted suggestions also frequently continue into the future.

All Federal agencies and corporations should tap the wellspring of employee creativity. No law or regulation requires establishing suggestion programs. However, Federal agencies and corporations have learned that suggestion programs are an effective way to harness employee creativity to improve operations and to save money.*

---

* Interagency Advisory Group Committee on Performance Management and Recognition Suggestion Program Working Group, *Good Ideas: A User's Guide to Successful Suggestion Programs* (N.P., 1995).

An interesting industrial practice from Japan combines the suggestion system with Kaizen in the Kaizen-Oriented Suggestion Systems (KOSS). Although national culture is, to some extent, the reason for Kaizen's success in Japan, a successful transfer of KOSS is less dependent on an amenable national culture than on the organizational culture. Kaizen activities in Japan involve everyone in a company—managers and workers—in a totally systemic and integrated effort toward improving performance at every level. Kaizen is a system that involves every employee, from upper management to the cleaning crew. Everyone is encouraged to come up with small improvement suggestions on a regular basis. This is not a once-a-month or once-a-year activity. It is continuous. Japanese companies, such as Toyota and Canon, receive a total of 60–70 suggestions per employee per year.

We should also distinguish between the employee suggestion system and *crowdsourcing*. Crowdsourcing is the practice of obtaining needed services or ideas by soliciting contributions from a large group of people, especially the online community, rather than from traditional employees or suppliers. Often used to fundraise for start-up companies and charities, this process can occur both online and offline. It sources valuable ideas and inventions from outside the walls of an organization. Not only does it bring more brainpower to bear on a problem to be solved, it also brings minds that are not constrained by industry conventions. Most recently, in April 2013 crowdsourcing helped to find the two Boston Marathon terrorists.

In summary then, the integration of value management with employee suggestion system will enable employees to be effective intrapreneurs (a word coined by Gifford Pinchot) that can deliver value to customers, think differently, and provide them opportunities to self-actualize and achieve their fullest potential.

19

The Value Educator Approach

The value of an organization consists first and foremost of the knowledge and skills possessed by the individuals who comprise it. The rate at which organizations learn may become the only sustainable source of competitive advantage. The finest organization is no better than the people who operate it. When you have the best people, you have the best product or service. The executive search firm Heidrick and Struggles has created a Global Talent Index that ranks countries according to their supply of talent. In their 2007 annual report they state, "Talent is the new oil and just like oil, demand far outstrips supply."

The organization that can attract, develop, and retain higher-caliber individuals guarantees the continuous replenishment of its knowledge base, ensuring a continuous increase in the organization's value-creating capacity and, therefore, its performance.

In organizations that have Value Services Departments or an Office of Operational Excellence, the value specialist becomes the Value Educator. When such an internal educator is not available, organizations hire management-consulting firms that specialize in Value Management to conduct value training for their employees.

In the United States, SAVE International is devoted to the advancement and promotion of VA/VE. SAVE International offers a comprehensive educational program in VA/VE, including a certification program. Some of the other professional societies in value management include the Canadian Society of Value Analysis, Society of Japanese Value Engineers, Chinese Society of Value Engineering, Indian Value Engineering Society, Hong Kong Institute of Value Management, Institute of Value Management-Australia, Institute of Value Management-UK, and the European Governing Board of the Value Management Training & Certification System.

These professional societies promote the following five commonly accepted elements of professionalism:

1. A unique body of knowledge
2. Rigorous standards of competency
3. Continued education to ensure currency
4. A potent ethical code of conduct
5. Dedication of service to society

The term *demonstration effect* refers to a peaceful political struggle in one place that may act as a catalyst for a peaceful struggle in another place. The success of the popular uprising in Tunisia that followed the fall of dictator Ben Ali in 2011 became the primary demonstration effect for Egyptian activists to overthrow the dictator Hosni Mubarak. Likewise, the success stories I cited from the U.S. Federal Highway Administration and the U.S. Army Corps of Engineers are other examples of the demonstration effect and serve to make a compelling case that *governments too can work better and cost less* by mandating the practice of VA/VE and leveraging its power as a strategic weapon.

If you want to know how a country can successfully move its young people from **education to employment**, think of the demonstration effect practiced in Canada. McGill University in Montreal pioneered an innovative course designed to give university students a chance to apply VA/VE by working with an industry participant on a real-world problem. The program has been in practice for the past 38 years, and has facilitated over 200 projects sponsored by over a hundred companies. Students work on an engineering problem provided by a company, involving product design, manufacturing processes, or service improvement. The program allows senior mechanical engineering students and industry-leading companies to become fully immersed in the systematic problem solving technique of VA/VE.*

If we want to change outcome, we need to change behavior, and to do that we need to change how we think systematically. That is why we need to ground students with the scientific method of value thinking before they enter the workforce. Developing a modern world-class workforce takes time as stated in the Chinese aphorism, "If you want prosperity for

---

* See http://www.mcgill.ca/ve/value-engineering.

one year, grow grain; if you want prosperity for ten years, grow trees; if you want prosperity for 100 years, grow people."

VA/VE is an ideal educational program that can help you blend classroom training with application to the job to provide contextual learning and thereby achieve results-driven organic rather than mechanistic change.

When a company unleashes the talent of its value-trained employees, it can self-diagnose its problems and self-discover its solutions. To that end, AOK Consulting & Education* conducts *blended learning* to promote organic change through a Value Improvement Event, "VIE Blitz," a highly focused, action-oriented 20-hour rapid-improvement pilot project. In this program an improvement team learns and applies the VA/VE scientific method of PISERIA, improves a specific process or product, and takes immediate action to implement the change, thereby achieving VA/VE's objective of creating a workforce of problem-solvers and intrapreneurs. Hence, if your investment in a high school or college education has not paid off for you, try to "learn-and-apply" this reengineered VA/VE.

As a value educator I espouse $P = A \times M$ as the formula for success. P stands for human Performance; A stands for Ability (knowledge and skills); and M stands for Motivation. This formula means that the success or performance of any organization is limited by the abilities and motivation of the people operating in it. Hence, the subtitle of this book, "VA/VE is the blueprint for developing a workforce of problem-solvers and innovators."

---

* See www.aokconsulting.com.

20

The Value-Organization Approach

VA/VE should not be treated as a new flavor-of-the-month initiative, but a foundation for everything your organization does by creating a culture of achievement and raising your organization to a higher level of excellence to maximize revenue, minimize cost, and achieve optimization.

VA/VE-oriented organizations practice value-minded management of their resources. The value program requires every function's involvement and top management's support. In the value-organization approach, successful organizations effectively integrate various analytical disciplines so that the whole becomes greater than the sum of the parts. Furthermore, effective value management requires that the whole business be reviewed. Otherwise, costs will be reduced in one place by simply being pushed to somewhere else.

What organizations need is to make value sensitivity contagious among their employees—an everybody-does-it approach—and thereby build creative and innovative organizations. To accomplish this goal, organizations should provide opportunities for value ability, value climate, and value experience.

VA/VE initiatives are focused on adding value for customers so businesses can grow. The key to achieving that success is for organizations to secure senior management support and adopt the following guideline to the fullest extent possible:

1. Locate sponsors, champions, and project managers—key influencers. Establish senior-level steering committee. Secure C-suite support.
2. Seek executive champions who will provide the necessary resources, remove any barriers, and communicate progress to higher management.
3. Invite participation by users and process owners.

4. Use workshop presentations and publicity at all levels of business—educate and inform. Promote **P**ersonal **R**esponsibility **I**n **D**aily Effort.
5. Demonstrate with quick wins (low-hanging fruits). Duplicate efforts.
6. Give your initiative an identity, such as SMX (Supply Management Excellence).
7. Promote teamwork and use functional expertise.
8. Win support with results, not promises. Walk the walk.
9. Drive the opportunity; don't wait for it to come to you.
10. The measure that management uses is financial performance, so talk their language to get their approval for Value Improvement Initiatives.
11. Demonstrate the ability of VA/VE to contribute to customer satisfaction.
12. Show contributions to strategic objectives. Commitment and involvement by executive leadership are vital.

A learning organization continuously expands its capacity to create its future.

21

Conclusion

Value Analysis & Value Engineering (VA/VE), a management tool created in 1947 by a U.S. business organization (General Electric) and then adopted in 1954 by the government (U.S. Department of Defense), has become one of the secrets of America's success and has been widely adopted by the industrialized societies.

The value of VA/VE is embedded in its systematic and disciplined methodology that focuses on function to meet or exceed customer requirements as well as identify and eliminate unnecessary costs that do not meet customer requirements. VA/VE is a better way to improve how we improve; it is how to do research work; it is about how to think systematically; it is about capacity building in analytical thinking, problem solving, and decision-making.

As a catalyst for change, the reengineered VA/VE updates and upgrades the value methodology by amalgamating the complementary strengths of Six Sigma, Lean Thinking, Business Process Reengineering, Kaizen, Total Quality Management, and Project Management into a value-improvement architecture that lets the steps of the value methodology (PISERIA) serve as the building blocks leading to optimization and operational excellence.

As you can see, VA/VE is not rocket science—it is common sense. But common sense is not common, which is why we need to own it and internalize it. Culture being how organizations do things, I am hopeful that you will adopt VA/VE as your corporate DNA—a way of doing things *better, faster, and leaner*—as it is worthy of being ingrained in your corporate culture, keeping in mind that "culture eats strategy for breakfast," as my great teacher Peter Drucker used to say.

If you are engaged in the pursuit of excellence and wish to make the leap from good to great and generate an immediate and measurable payback to the tune of at least 20:1 (that is 2000% return) for every dollar invested in your value-improvement effort, I urge you to dare to be great and earn your bragging rights as a master value specialist by practicing VA/VE and achieving operational excellence.

You are not defined by your past; you are prepared by your past. Today is the first day of the rest of your business life. If you are value-minded you think differently. By adopting VA/VE as your template for success, you now have the greatest potential to make a difference and live a life of excellence! Since every job is a self-portrait of the person who did it, start autographing your work with excellence and enjoy the happiness that comes from that! Embrace the power of one—the one value education that will help you make a difference.

Wisdom is knowing what to do; **skill** is knowing how to do it. **Success** is doing it *now*. I hope you are fired up and ready to go value hunting! At work or in school, value sensitivity could be contagious—let's start an epidemic! I measure my success by yours. So, when you achieve results with VA/VE please let me know your success story by using the Value-Improvement Proposal form in Appendix C. You can reach me via e-mail at aok@aokconsulting.com or kassa.abate@gmail.com.

In the world of VA/VE, you don't buy a book; what you buy is what the book will do for you. I hope you have found my book to be worth your investment or found it to be good value-for-money!

I would like to take this opportunity to pay tribute to the VA/VE trail-blazers on whose shoulders we stand, Larry Miles and Carlos Fallon, with whom I had the honor to be pictured at a 1972 SAVE International annual conference in Miami, Florida.

Appendix A: The Value Process Dashboard

The Value Process				
Value Job Plan	**Objectives**	**Key Questions**	**Specific Techniques**	**General Techniques**
1 PREPARATION	Identify/recognize/ observe problems & opportunities Select value work Select value worker(s) Schedule the effort	What are the right problems that we should be solving?	Use project selection criteria. Separate & set priorities & posteriorities. Apply Pareto's Law to establish priorities. Relate goals to company objectives. Follow the value job plan.	
2 INFORMATION	Define problem Define functional requirements Obtain background information Analyze data	What is it? What does it do? What must it do? Which is the basic function? What does it cost? What is it worth?	Identify the item by name, material, size, shape, etc. Identify the function. Use two-word definition. Determine basic and secondary functions (also by functional areas). Learn marketing requirements. Gather facts concerning usage, specifications, material, manufacturing methods, and processes: get drawings and samples. Separate facts from opinions. Appraise information for relevance, freshness, and accuracy. Avoid the generalities trap—be specific. Use charting to collect and present data. Review cost data—material, labor, and overhead.	Work on specifics Get information from the best source
3 SPECULATION	Create new ideas Form the hypothesis	What else will perform the function?	Create climate for new ideas. List all ideas offered—no judging. Use brainstorming, synectics, checklisting. Consider alternate and speciality products, processes, and materials. Eliminate, combine, substitute, standardize, simplify, change (place, sequence, person).	Use good business judgment
4 EVALUATION	Select best ideas Develop the best alternative Test the hypothesis (experiment)	What are the best ideas? What will these ideas cost? What is the best alternative? What will this alternative cost? What is it worth?	Screen ideas for possibles, probables, and impossibles. Evaluate by comparison. T-chart good and bad features. Put dollar signs on key tolerances. Consider other functions (work or sell) and specification features that must be incorporated. Verify technical feasibility (make actual models), customer appeal, and savings. Refine ideas that show promise of providing improved value. Consult suppliers, industry and company specialists.	Inspire teamwork
5 RECOMMENDATION	Prepare proposal Make presentation	How should this alternative be presented? What is required to sell the proposal? Who has to OK it?	Prepare visual aids. Specify plan of action for implementation (if possible). Motivate positive action.	Overcome roadblocks
6 IMPLEMENTATION	Follow implementation plan Install the new method	Are engineering change orders prepared?	Obtain skills, facilities, resources. Obtain improvement and expenditure authorizations.	Use good human relations
7 AUDIT Results	Follow-up Audit actual results Report innovation	What was done? — after phase one? — after recommendation? What was the result? — after implementation?	Document VA savings. Release periodic VIP report. Publicize success stories.	

Appendix B: Value Stream Plan (What to Do, by When)

PROJECT PLANNING & CONTROL

Value Improvement Process

PROJECT No.____

PROJECT DESCRIPTION: _____

Team Chairman: _____
Team Secretary: _____
Other Members: _____

PROJECT SAVINGS _____
PROJECT EXPENDITURE _____ =
Other Benefits (attach memo) _____

DATE STARTED: _____ COMPLETED: _____

Advisor (if any): _____

	PROJECT STEPS	Week of	Week 1 M T W T F	Week 2 M T W T F	Week 3 M T W T F	Week 4 M T W T F	Week 5 M T W T F	Week 6 M T W T F	Week 7 M T W T F	Week 8 M T W T F
PRE FEASIBILITY — PREPARATION	Identify project candidates									
	Select priority project									
	Define scope and objective									
	Determine project worth									
	Select project worker(s)									
	Approve project start									
	Schedule the value effort									
FEASIBILITY STUDY — INFO	Collect data									
	Analyze data									
FEASIBILITY STUDY — SPECULATION	Create new ideas—no judging									
	Apply brainstorming, synectics, checklisting; eliminate, combine, substitute, standardize, simplify, etc.									
FEASIBILITY STUDY — EVALUATION	Develop evaluative criteria									
	Analyze to select best idea(s)									
	Develop best value alternative									
	Consult specialists									
PROJECT PROPOSAL — RECOM-MENDATION	Prepare proposal									
	Make presentation									

Appendix C: Value Improvement Proposal

In submitting your proposal, please follow these steps:	DAY	MONTH	YEAR	PROJECT NO.
1. DESCRIBE THE PRESENT METHOD OR PROBLEM. (What is it? What does it do? What does it cost?)				
2. EXPLAIN YOUR PROPOSED METHOD OR SOLUTION. (What else will do? What will that cost?)				
3. POINT OUT THE SAVINGS AND/OR BENEFITS THAT CAN RESULT FROM YOUR PROPOSAL	PROJECT TITLE			

_____(Add more space here) _____

ESTIMATED COST SAVINGS SUMMARY

COST FACTORS	PRESENT	PROPOSED	SAVINGS	% REDUCTION
Material Cost per unit_____				
Labor Cost per unit_____				
Overhead Cost per unit_____				
TOTAL COST per unit_____				
Annual Requirements_____				
TOTAL ANNUAL COST_____				
GROSS SAVINGS FIRST YEAR _____				
Less Implementation Cost				
Cost of Value Study_____				
NET SAVINGS FIRST YEAR_____				

Additional Benefits (Please use following page of this form)_____

APPROVED BY_____ DATE_____

Engineering Change Order No. (Implementation Letter) _____

PROJECT WORKER (s) _____

Appendix D: Evaluation Report and Guide

D-1: EVALUATION REPORT

YOUR company name and address _____ Evaluation Report__

Project No._____ Project Title_____
Project Worker (s)_____

A. Please check (x) the appropriate boxes below:
    1. __ Tangible Suggestion/Proposal.   __ Intangible Suggestion/Proposal.
    2. This Suggestion/Proposal requires contacting the suggestor for clarification or additional information. YES__ NO __
    3. This idea can be used locally __ and also by other departments system wide __.
    4. This idea was __ was not __ under consideration by the management of this department and/or division
       prior to the receipt of this Suggestion/Proposal.
    5. This Suggestion/Proposal will be adopted in part __ will be adopted in its entirety __ will not be adopted __.
    6. This Suggestion/Proposal will:
       __ Increase Productivity __ Reduce Cost   __ Combine Operations   __ Save Time __ Improve Service
       __ Prevent Accidents __ Eliminate Waste __ Simplify Methods __ Improve Quality __ Improve Working Conditions

B. EVALUATOR'S RECOMMENDATIONS (Please use the Evaluation Guide on the next page for savings and benefits
    appraisal. Please give specific reasons for adoption, non-adoption or deferment.)
    _____
    _____
    _____
    _____

C. IMPROVEMENT AUTHORIZATION (If this is an employee suggestion, the benefiting department will be the resolving
    office. If this proposal is acceptable and it is within your normal authority to adopt or to initiate the necessary action, please
    do so. Then describe here the action taken, including dates and references to implementation authorization documents.)
    _____

    Implementation Date _____ Reference _____
    Appraised By _____ Position _____ Date _____
    Approved By _____ Position _____ Date _____

D-2: EVALUATION GUIDE

EVALUATION GUIDE

1. TANGIBLE SAVINGS COMPUTATION

COST FACTORS	Present	Proposed	Savings	% Reduction
Material Cost per unit _____	_____	_____	_____	_____
Labor Cost per unit _____	_____	_____	_____	_____
Other Costs per unit _____	_____	_____	_____	_____
TOTAL COST per unit _____	_____	_____	_____	_____
Annual Requirements:_____				
TOTAL ANNUAL COST	_____	_____	_____	_____

Gross Savings First Year _____
Less Implementation Cost (amortized) _____
Net Savings First Year _____
TANGIBLE SUGGESTION AWARD = Net First Year Savings x _____% = $_____

2. INTANGIBLE BENEFITS APPRAISAL (Quality, Safety, Working Conditions)

FACTORS	Very Low	Low	Average	High	Very High
EFFECTIVENESS (How effectively does this suggestion correct the situation?)	0	10	20	30	40
SERIOUSNESS OF PROBLEMS OR DEGREE OF HAZARD (How undesirable is the condition that this suggestion seeks to improve?)	0	6	12	18	24
INGGENUITY AND COMPLETENESS OF SUGGESTION (How much initiative, imagination, or ingenuity is shown?)	0	4	8	12	16
EXTENT OF APPLICATION (What portion of employees or what extent of operation is affected?)	0	3	6	9	12
FREQUENCY OF OCCURRENCE (What are the chances of the conditions occurring?)	0	2	4	6	8
					100

Total Points _____ x $_____ = $_____ Less Implementation Cost _____
INTANGIBLE SUGGESTION AWARD: _____
AWARD AUTHORIZATION _____

Appendix E: Preparation Agenda

VA/VE's Scientific Method of PISERIA

1. Preparation
Identify project candidates

Perform Spend Analysis
Perform SWOT Analysis
Perform Pareto Analysis
Select priority project/triaging
Define objective and scope
Define project deliverables
Determine project worth
Select project worker(s)
Complete Project Charter
Approve project start
Schedule the value effort

2. Information
Define problem,
Analyze data,
Identify root cause

3. Speculation
Generate
potential
solutions

7. Audit Results
Review, report, act,
institutionalize change

6. Implementation
Introduce change/
solution

5. Recommendation
Propose change/
solution

4. Evaluation
Evaluate/
select solution

Appendix F: Information Agenda

VA/VE's Scientific Method of PISERIA

1. Preparation
Identify project candidates

7. Audit Results
Review, report, act,
institutionalize change

Problem definition

Collect and analyze data

Value stream mapping

Identify root cause

Identify constraints

Symptoms vs. problems

Tools: Fishbone
  5 Why's
  QFD

2. Information
Define problem,
Analyze problem,
Identify root cause

3. Speculation
Generate
potential
solutions

6. Implementation
Introduce change/
solution

5. Recommendation
Propose change/
solution

4. Evaluation
Evaluate/
select solution

Appendix G: Speculation Agenda

VA/VE's Scientific Method of PISERIA

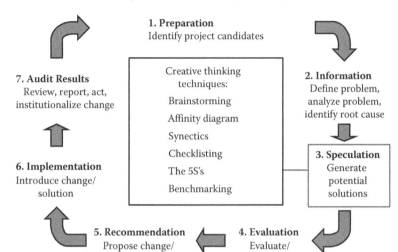

1. Preparation
Identify project candidates

7. Audit Results
Review, report, act,
institutionalize change

Creative thinking
techniques:
Brainstorming
Affinity diagram
Synectics
Checklisting
The 5S's
Benchmarking

2. Information
Define problem,
analyze problem,
identify root cause

6. Implementation
Introduce change/
solution

3. Speculation
Generate
potential
solutions

5. Recommendation
Propose change/
solution

4. Evaluation
Evaluate/
select solution

Appendix H: Evaluation Agenda

VA/VE's Scientific Method of PISERIA

1. Preparation
Identify project candidates

7. Audit Results
Review, report, act,
institutionalize change

Develop evaluative criteria

Analyze to select best idea(s)

Forcefield analysis

Develop best value alternative

Risk analysis

Decision analysis

Consult specialists

2. Information
Define problem,
Analyze problem,
Identify root cause

3. Speculation
Generate
potential
solutions

6. Implementation
Introduce change/
solution

5. Recommendation
Propose change/
solution

4. Evaluation
Evaluate/
select solution

Appendix I:
Recommendation Agenda

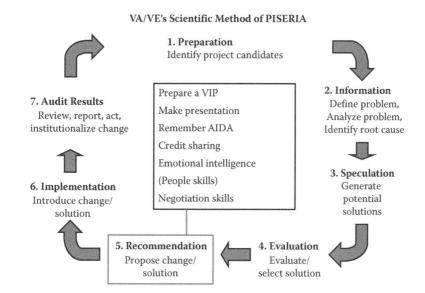

VA/VE's Scientific Method of PISERIA

1. Preparation
Identify project candidates

7. Audit Results
Review, report, act,
institutionalize change

Prepare a VIP
Make presentation
Remember AIDA
Credit sharing
Emotional intelligence
(People skills)
Negotiation skills

2. Information
Define problem,
Analyze problem,
Identify root cause

3. Speculation
Generate
potential
solutions

6. Implementation
Introduce change/
solution

5. Recommendation
Propose change/
solution

4. Evaluation
Evaluate/
select solution

Appendix J:
Implementation Agenda

VA/VE's Scientific Method of PISERIA

1. Preparation
Identify project candidates

7. Audit Results
Review, report, act,
institutionalize change

Creativity is not enough

Implementation enablers

Implementation barriers

Implementation plan

Pilot plan

Execute plan

2. Information
Define problem,
Analyze problem,
Identify root cause

3. Speculation
Generate
potential
solutions

6. Implementation
Introduce change/
solution

5. Recommendation
Propose change/
solution

4. Evaluation
Evaluate/
select solution

Appendix K:
Audit Results Agenda

VA/VE's Scientific Method of PISERIA

1. Preparation
Identify project candidates

7. Audit Results Review, report, act, institutionalize change	**2. Information** Define problem, Analyze problem, Identify root cause

Ideas–practice–feedback

Verification

Achieve prior "buy-in"

Document lessons learned

Institutionalize best practice

Recognition/celebration

Formally close project

3. Speculation
Generate
potential
solutions

6. Implementation
Introduce change/
solution

5. Recommendation
Propose change/
solution

4. Evaluation
Evaluate/
select solution

Author

Abate O. Kassa, author, consultant, and educator in Value Management, is president of AOK Consulting & Education, established in 1973 and based in New York. He is a former purchasing manager for Ethiopian Airlines, where he introduced a corporate-wide value-improvement process after value-training 343 management staff and also revitalized its employee suggestion system. Mr. Kassa combines resource and process consultancy to provide organic change for his client organizations instead of the traditional mechanistic change. As a Nichepreneur™ specializing in value management and purchasing & supply management, he assists his clients to optimize the value of their operations.

Mr. Kassa has also gained experience in international public procurement working with the International Trade Centre/UNCTAD/WTO on a $1 billion import procurement project in Africa and also served as an ITC consultant in the Pacific Forum Island Countries of Samoa, Kiribati, Tonga, Niue, Tuvalu, and Fiji. He was a regular speaker for many years at the affiliates of the Institute for Supply Management and the American Management Association in the United States, where he used his own workbooks (unpublished) to teach Value Management and Purchasing & Supply Management.

Mr. Kassa was retained as executive director of the Institute for Supply Management-New York for 19 years (1981–2000). He was the recipient of ISM-New York's 1992 J. H. Leonard Award for his contributions in institutional capacity building of ISM-New York as a center of excellence in purchasing and supply management.

Once a science master at his elementary school as a teenager, he has now anchored the value methodology in the common knowledge of the scientific method to reengineer value engineering; and this book is the product of his dual lenses of experience and research over many years.

Mr. Kassa is a lifetime Certified Purchasing Manager (CPM) with the Institute for Supply Management and a former Certified Value Specialist (CVS) with SAVE International. He holds a bachelor's degree in economics from Addis Ababa University, Ethiopia, and a master's degree in government and politics from St. John's University, New York City, New York.

Index